Wintering in Snow Country

Wintering in

by William Osgood and the

THE STEPHEN GREENE PRESS

Snow Country

Editors of The Stephen Greene Press

Drawings by Edwin E. Fletcher, Jr.

Brattleboro, Vermont

ACKNOWLEDGMENTS

The author and editors wish to thank the following helpful advisors—all of them winterers in snow country, former or present. They have added much to the value of this book, but are in no way responsible for its shortcomings: Randy Cross, Wilmington, Vermont; Robert L. Greene, Wilmington, Vermont; Walter Hard, Jr., Burlington, Vermont; J. William Hasskarl, Brattleboro, Vermont; William Huestis, Huestis Supply Company, Inc., Brattleboro, Vermont; Rod Linz and Henry E. Merrill, Barrows 35 company, Brattleboro, Vermont; Francis Rowsome, Jr., Washington, D.C.

This book has been produced in the United States of America.
It is published by The Stephen Greene Press,
Brattleboro, Vermont 05301.

LIBRARY OF CONGRESS CATALOGING IN PUBLICATION DATA

Osgood, William E.   1926–
    Wintering in snow country.
    Includes bibliographies and index.
    1. Home economics. 2. Winter. 3. Dwellings
—Maintenance and repair. 4. Automobile driving
in winter. I. Stephen Greene Press. II. Title
TX147.083   640   75–8193
ISBN 0-8289-0319-0
ISBN 0-8289-0320-4 pbk

PUBLISHED APRIL 1978
Second printing August 1978

Contents

1

Wintering in Snow Country

ALL OF CANADA and much of the United States above the 42nd Parallel of latitude are covered in winter with a continuous mantle of snow that remains throughout most of the winter season from the Winter Solstice on December 22 to the Spring Equinox on March 21. In much of this snowbelt winter may stretch well beyond these dates, with snow and cold weather lasting from October to May. Especially for rural dwellers, winter in snow country is a season to be reckoned with, prepared for, endured, and—ideally—enjoyed.

Of course, no definition of the "Snowbelt" as an area north of a certain line can be entirely satisfactory: there is plenty of snow in the Arizona mountains, and winters in Vancouver are relatively mild. Exceptional instances of winter weather that seem geographically improbable are many, and are produced by complex sets of local influences. In spite of exceptions, however, the rough demarcation of the snowbelt as the northern two-fifths of the United States plus Canada, stands—it is here that winter cold and snow are a major fact of everyone's life, to be anticipated and coped with.

Winter cold generally gets more intense (again, with some exceptions) as the latitude increases towards the poles. From the time of the Fall Equinox (September 22) until the Winter Solstice the Northern Hemisphere turns away from the area warmed by the sun's radiation. As a result the days get shorter and shorter the farther you travel towards the North Pole in this winter period, and mean temperatures get lower and lower—because the farther north from the Equator, the farther the hemisphere tilts

1

away from the sun's warmth. Hours of darkness are increased until you reach the point where the sun doesn't rise above the horizon at all.

LOCAL WEATHER-MAKERS

Despite the general trend towards lower temperatures with distance from the Equator, high latitude alone does not always produce a bitter winter season. Ocean currents like the Gulf Stream carry huge amounts of warm water along the coasts of the continents, and this warm water tempers the air enough to make the winters comparatively mild even in high latitudes. Therefore the west coasts of Norway and of the United States have rather mild winters for all their northern situations, and Iceland, at the edge of the so-called North Atlantic Drift—an extension of the Gulf Stream—is surprisingly warm despite its name.

Large bodies of water can moderate winter temperatures, but when water has turned to ice the scene changes dramatically. Winds blowing over large masses of ice can bring a real spell of prolonged cold weather. People living in the northern part of the Province of Quebec can tell almost to the day when Hudson Bay has frozen over completely; for while the bay is open their weather is cloudy and snowy, but when the ice cover is complete or nearly so they can expect extended periods of clear days with very low temperatures.

The local influence of water on winter weather has been described as a lake effect, and one of the best places to observe the effect at work is in the area of the United States and Canada just east of Lake Ontario. This lake has the smallest surface area of all the Great Lakes, while at the same time it is the second deepest. This relationship of depth to area results in Lake Ontario's having a very large heat-storage capacity. Even in the severest of winters only about a quarter of the lake freezes over; normally only fifteen percent has an ice cover. Prevailing westerly winter winds blowing over the open lake become very unstable and also pick up moisture. As the winds are forced upwards over the highlands on the lee side of the lake, the air is cooled and much of the moisture they picked up from the lake is precipitated out, to make this the snowiest zone east of the Rocky Mountains.

In order to develop a lake effect of significant magnitude, a body of water must have large heat-storage capacity. Lake Ontario is perfectly built and conveniently situated for generating this phenomenon. On the other hand, Lake Erie, just to the west, is quite shallow. Over one-half of its surface is covered with ice in a mild winter and normally the lake freezes over almost entirely. Snow removal crews in Buffalo, New York, and adjacent towns and cities get some rest after Lake Erie freezes up, because an ice-covered body of water can no longer contribute abundant moisture to the passing winds. Lake effects, great and small, like those associated with lakes Erie and Ontario, are tremendously important to a study of winter in general and snow zones in particular.

Frozen bodies of water prolong winter beyond the three months between the Winter Solstice and the Spring Equinox. Spring can never really take over from winter until ice has turned to water, and frozen water requires vast quantities of the sun's heat to melt. Although winter is extended by this phenomenon to reach out into the domain of spring, it can also be said that the onset of winter is delayed by water bodies, which cool at a slower rate than the surrounding land. Savings of heat are put on deposit during the summer when fields, forests, lakes and streams and the great oceans absorb solar radiation. Then, as parts of the globe turn away from the sun, this heat is paid out at varying rates. Land pays out its heat deposits rather quickly, while water is slower to cool.

Wind currents also affect the severity of winters in different places. My own state of Vermont, a place of long, cold winters, is on about the same latitude as southern France, where winter as a cold season is practically unknown. Vermont, New Hampshire and Maine, plus the neighboring Canadian border towns, have such severe winters, despite the fact that they are only about mid-way from the Equator to the North Pole, because vast amounts of cold polar air sweep over them carried by the prevailing westerly winds. Still, dry air is a marvelous insulator of heat; but get it moving in the form of wind and then air can dissipate great quantities of heat to make middle-latitude spots like Vermont into cold winter paradises for skiing and other snow and ice sports.

Once the great bulk of heat, including that stored in lakes, ponds and streams, has been dispersed into outer space, then winter can really assert itself. This forms the basis of the old folk saying, "As the days begin to lengthen, the cold begins to strengthen." That is, although more sunlight reaches earth as the days grow longer from late December to March, mean temperatures drop because earth and water by then have given up the heat they absorbed in summer.

When the earth has cooled enough that fallen snow doesn't melt, another factor assures the winter regime of a firm grip. This factor is the **albedo,** or reflective power, of fallen snow. White surfaces reflect heat and light. Clean snow can reflect as much as ninety percent of solar energy. Therefore, when the sun does shine during the slowly lengthening days after the Winter Solstice, much of the heat is reflected right back into space. At night, especially if there is a minimum of cloud cover, the earth continues to lose heat through radiation, which helps to account for the extremely low temperatures on cloudless winter nights.

Another considerable influence in causing local winter conditions is **elevation** above sea level. Under normal conditions one can expect a lowering of the thermometer by three degrees for each one-thousand-foot increase in elevation in the envelope of atmosphere nearest the earth. There are exceptions to this rule, however, an important one of which is that under certain weather conditions the thermometer may in fact be several degrees warmer at higher elevations in winter: the really cold spots are in the valleys. These reversals of the standard temperature lapse rate—**temperature inversions**—take place quite frequently on still nights when the heavy, cold air slides down from the hills to the valleys, and warm air rises to the higher elevations.

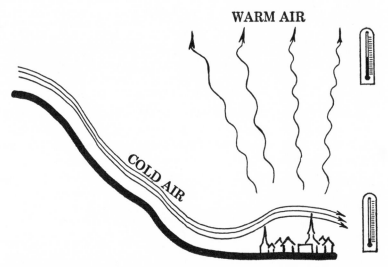

Temperature inversion: warm air rises to high ground as colder air descends, producing lower temperatures in low-lying areas than on heights.

SNOW

Cold weather clouds are made up of ice crystals and minute water droplets which remain in liquid form even though the air temperature is below freezing. These "supercooled" droplets and the ice crystals themselves are individually formed around a central nucleus which may be a tiny particle of dust or of sea salt. It's quite remarkable how much sea salt is held in suspension in the atmosphere. Since salt has the ability to draw water to itself, it is the most common nucleus for the supercooled water droplets in clouds.

As long as a cloud remains in a stable condition nothing falls out of it, but if conditions change in one way or another a snowstorm can result. When, for example, a cloud rises to a higher altitude in passing over a mountain range, the colder temperatures at the high altitude cause its water to freeze and adhere to the ice crystals, which get larger and heavier until they are no longer able to float in the air. The crystals then fall to earth, as snow or as one of the three types of frozen precipitation other than snow: graupel, ice pellets and hail (which last is associated with summer thunderstorms and is not a winter phenomenon).

Ice crystals that fall as snow arrive on earth as snowflakes—aggregates of crystals which have become attached to each other en route to the ground. What we see in the air are primarily snowflakes, and not the individual crystals as they were born in the clouds.

Ice pellets (sleet) resemble hailstones in their rounded shape, and do not obviously exhibit the distinctive crystal designs. But while hail occurs in the summer, ice pellets fall in the winter. They are often referred to as frozen rain.

Graupel is formed when supercooled water droplets attach themselves as a frozen coating (rime) onto ice crystals in clouds. Graupel is actually heavily rimed snow crystals, also referred to as snow pellets, but they are easily distinguished from ice pellets because graupel has a light shade whereas ice pellets are dark. When graupel strikes a hard surface it bounces—you can hear it fall. Within the snow cover graupel tends to keep its individual form for long periods of time. It makes a good base for ski touring over gently rolling terrain.

WINTER STORMS AND STORM WARNINGS

Weather scientists have found that North American winter storms have rather definite breeding grounds, and that these areas are the source of most of the important storms that significantly affect human activities. In the Pacific zone, strong low pressure systems build up off the east coast of Asia and then travel towards Alaska where they leave great quantities of snow in the southern part of the state. Other Pacific storms originate in mid-ocean and travel easterly to strike the western United States and Canadian provinces. The great bulk of this snow falls on the western slopes of the mountain ranges nearest the sea, including the Sierra Nevada and the Cascades. Record-breaking snow depths often occur here.

Most Pacific Ocean storms of this type dissolve over the mountain ranges, but from time to time the storms may reconstruct themselves and start out with renewed vigor to travel eastward across the continent. This storm redevelopment pattern seems to concentrate over the Rocky Mountains in Colorado and Alberta. These storms often move east, homing on the Great Lakes, where their intensity is often accentuated by the well-known lake effect. Passing the lakes, the storms may dump heavy snow in the area from Buffalo to Watertown, New York, and then continue on down the St. Lawrence River valley, dropping more snow on central and northern Quebec and Newfoundland before the storms finally break up in the North Atlantic south of Greenland. These zonal flow storms are fairly easy to predict, and usually enough advance warning can be given so people can prepare for them.

Another major source of winter storms is off the Carolina and Virginia coasts out in the Atlantic, or inland east of the Appalachian Mountains. This is the birthplace of those violent storms called **Northeasters,** which move up the Atlantic coast often affecting inland regions to a considerable degree. New Englanders and folks in the Maritime Provinces have a healthy respect for Northeasters.

The United States National Weather Service and Canada's Meteorological Branch of the Department of Transport take great pains to pinpoint storm breeding areas and to watch them closely for indications of activity. After a storm develops, weather offices located strategically around the country track it (assisted nowadays by earth satellites) and issue advisories to the areas that might be affected. Two sorts of advance notice are given. The first is called a *winter storm watch,* or advisory, meaning that a region may be affected. If it seems quite certain that a significant storm will then descend on the region the watch advisory is changed to a *warning.*

Most snowstorms will deposit up to four inches of snow over several hours. Falling rates of an inch per hour or more are considered to be heavy. In regions where heavy snowstorms are rare a storm that may be expected to leave only two or three inches might be warned in advance as a heavy storm. This would also apply to metropolitan areas where traffic tie-ups can be generated by what some might think of as a gentle storm. In other parts of the country, where storms leave more abundant snow, a heavy snow warning would be issued if it was thought that at least six inches might fall. Except in coastal regions of the West, southern Alaska, east of the Great Lakes, and in the vicinity of the Gulf of St. Lawrence, where extremely heavy snowfalls can be expected much of the winter, most storms leave relatively small amounts of snow.

Winter storms have been classified according to severity, and it is agreed that the most severe of all is the **blizzard**. These violent storms consist of snow, high winds, and intense cold. They wreak terrible havoc out in the mid-continent plains where the winds can sweep unobstructed over the vast level or rolling landscape. The wind blows the snow with such ferocity that visibility is zero. People get lost right among familiar surroundings, and those who go to look for them get lost in turn.

Blizzard warnings are given when it is expected that winds will blow at least thirty-five miles per hour accompanied by heavy snowfall and temperatures of twenty degrees Fahrenheit or lower. Even more ominous is a **severe blizzard** warning, which means that heavy snowfall will be blown about with winds of forty-five miles per hour or stronger combined with temperature readings of ten degrees or less. Neither man nor beast should be abroad during a severe blizzard.

High winds are the most destructive aspect of blizzards. Because of the well-known effect of "wind chill," even a gentle breeze can lower the apparent temperature so that the air feels much colder than it is (See Appendix II for more on Wind-Chill). Violent winds also drift and blow the snow to keep visibility at a minimum even after the snow precipitation from the clouds has stopped.

Severe blizzards are fortunately exceptional; but minor storms can also be very dangerous. A particularly dangerous kind of storm is the glaze or **ice storm,** which occurs when rain falls in winter and freezes on contact with the earth's surface and onto vegetation, utility wires and all exposed objects. Roads and highways get so slick that it is impossible to maintain traffic: autos skid completely out of control even in the hands of the most expert drivers.

Ice storms put tremendous burdens on utility wires. If the wires themselves do not break under the weight of ice that collects on them, they are often severed by falling branches from nearby trees. Power failures are caused by ice storms at least as often as by more spectacular snowstorms and blizzards. Trees also suffer great damage. Ice storms are most frequent in the southernmost regions of the wintry zone, and it has often been remarked that trees are more shapely in the north because they have not been bent and ruined by the ice storms that occur to the south.

Sleet is close kin to ice storms, with the difference that sleet storms are made up of ice pellets while ice storms are composed of rain which immediately freezes on contact with cold surfaces. Ice pellets making up a sleet storm act just like graupel (snow pellets) by bouncing when they strike a hard surface.

In the gamut of winter storms there are two other important types that are not generally known because of their localized influence. These are **snowbursts** and **snow squalls,** which apparently are closely connected with the lake effect and are most often seen in Ontario and upper New York state. The term, snowburst, is applied to storms that approach blizzard proportions by virtue of having very heavy snowfall associated with high winds. The differences between snowbursts and blizzards seem to be that a snowburst may not have the low temperatures that always accompany a blizzard, and that snowbursts are relatively brief.

A snow **flurry** is the more gentle cousin of snowbursts and snow squalls. We see lots of snow flurries in the course of a winter. These mild storms do much to replenish the snow cover and keep it clean.

PREDICTING WEATHER

It seems that almost everyone wonders what the shape of the winter to come will be. A host of folklore weather forecasters may be consulted in predicting winter weather, including woolly bear caterpillars, muskrat houses, corn husks and chicken feathers. In fact, most of these have more to tell about the character of the growing season just past than about the winter ahead. All told, my own experience leads me to doubt the value of woolly bears and their associates as winter weather predictors.

Abe Weatherwise and his colleagues who make up the long-range forecasts for the almanacs base their predictions on past weather conditions, and probably take into consideration nowadays the vast store of climatological observations that have been accumulated by the National Weather Service. When you are planning some kind of winter excursion or camping trip, always take note of the daily local weather forecasts, which are very dependable in predicting conditions one or two days in advance. Long-range forecasts are less accurate. Weather maps, compiled by the Weather Service and published in many newspapers, are also helpful.

There are also some rules of thumb that can be used for amateur forecasting which I've found to be reasonably accurate in predicting local weather. Certain conditions seem to be quite faithful as short-term predictors of weather to come. I'd like to offer a few of these short-range weather signs for consideration.

Sky color is one sign that is quite reliable, as in the old verse, "Red sky in the morning, sailor take warning." To be more appropriate for snowbelt residents I have taken certain liberties with the verse, as follows: "Red sky in the morning, skier start waxing." This is not as euphonious as the original, but it is just as apt. If the red sky at sunrise is followed quickly by a change to lemon yellow, a storm followed by cold weather is often in store. The same lemon yellow hue often foretells a cold and clear night.

A halo around the sun or moon is another good sign of a coming storm. The halo itself is made up of snow crystals for the most part. An effect similar to the halo is the rarer phenomenon called a **Sun Dog** or *parhelion*—a bright spot like a second sun appearing on the horizon to one side of the sun itself. These beautiful winter sights are most frequently observed at sunrise, when there may be one sun dog on either side of the sun. At other times, when the sun is higher in the sky, a colored circle may go all the way around the sun. A related phenomenon to a sun dog, the *anthelion*, is a luminous white spot in the sky opposite the sun. If you see dogs or anthelia, get out the snow shovel.

The high thin cirrus clouds commonly called **mares' tails**, made up of tiny ice crystals, run in advance of storms by a day or two. Mares' tails seldom obscure the sun very much, but later in the time sequence preceding the actual storm, the clouds may thicken and cover the sky entirely. In mountainous or hilly country there is another weather sign which is quite accurate concerning the clouds. When the tops of the hills are in the clouds while the lower elevations are quite clear, a storm may be expected very soon—within a few hours. New Englanders say, "When the clouds are on the hills they will soon come by the mills." This version obviously was intended for summertime when the rains came to swell the brooks and rivers, but the principle holds for wintertime as well.

Other weather signs have to do with the way the temperature "feels" to the human body. The feeling was best expressed by the poet John Greenleaf Whittier in his "Snow Bound":

> *A chill no coat, however stout*
> *Of homespun stuff could quite shut out,*
> *A hard, dull bitterness of cold,*
> *That checked, mid-vein, the circling race*
> *Of life-blood in the sharpened face.*

This **shivery-cold feeling** results from high humidity. Although it feels very cold, the actual temperature reading on the thermometer may not be far below freezing. Most sizable snowstorms occur when the temperature is not too low, because very cold air cannot hold enough moisture to create a heavy snowfall.

Sounds are another good indication of the imminence of a storm. Sounds carry very distinctly and can be heard for long distances just before a storm. Sometimes when my next-door neighbor down the road is splitting wood, I listen for the clink of his hammer on the steel wedges and try to calculate by the loudness of the sound how far off a storm may be.

A **shift in the wind** direction also is one of the hints to look for in predicting a weather change at home. My weather vane was made especially for me by a friend. It is in the shape of a man on skis leading a reindeer, and I keep careful watch on my unique weather vane to see which way this pair is heading. In the old terminology the wind is said to be "veering" or "backing": wind which is moving against the apparent movement of the sun is backing, while wind which shifts along with sun movement is veering. A backing wind indicates a storm is coming.

Another sign, though I find it to be less reliable than the others I have mentioned, is the **hungriness of birds.** It seems to me that the birds have some way of knowing that bad weather is moving in, and they try to eat as much as possible before the snowstorm hampers their movements and covers their forage. So a bird-feeder can be added (with some reservations) to the homely array of instruments for the amateur weather forecaster.

When snowflakes begin falling out of the leaden skies it behooves one to take notice of the **size of the flakes,** as this also may be some indication of the magnitude of the storm. It has been my observation that if the flakes are comparatively small and coming down thickly there will be a considerable accumulation before the storm is over. On the other hand, if the flakes are large and fluffy, the storm will not leave as much snow. I'll note, however, that snowflake size as a portent of the size of the storm can be very deceptive. The types of snow and ice crystals in snowflakes often change quite markedly during the course of any individual storm, changing the size of the snowflakes themselves.

Not all weather signs are of ancient standing. **High jet trails** can be used to predict storms. If these vapor trails fade quickly the weather will hold fair, but if the trails hang in the sky for a long time, and enlarge into broad bands across the sky, a storm is apt to be in the offing.

LATE WINTER WEATHER

Each winter's snow cover usually develops with a fairly consistent pattern of rapid accumulation in the early part of the season followed by a leveling off in mid-winter, when the build-up of new snow added to the surface by storms is offset to a degree by internal changes in the snowpack itself. Processes going on within the snowpack, including settling, compaction, thawing and freezing, reduce the depth of the snow cover. It is often estimated that the depth of snow on the ground at any one time is only a third of the winter's total snowfall.

The snow cover may seem to be an inert white blanket, but in reality snow on the ground is undergoing constant change. Partial melting, differences in vapor pressure within the snow cover, the compressive force of the snow's own weight, may cause the snowmass to settle or move, may cause its internal structure to be looser or denser, wetter or drier. Even in the depths of winter, after the Winter Solstice, the snow cover continues to change its character; but with the advance of spring the changes become more drastic and their pace accelerates.

After the Solstice in late December the Northern Hemisphere begins to tilt back towards the sun: gradually the days lengthen. Although King Winter is never more firmly seated on his throne than during the weeks after the Solstice, the celestial spring, so to speak, has already begun. There are those of us who would prefer winter to taper off with gradually diminishing snow cover through late March and into April. This does happen over much of North America, although for New Englanders and some Canadians there is an unseasonable preview of April in the form of the January Thaw, which really is not of much help to anyone except in the way of saving some fuel. Most of the time a January thaw lasts only a day or so, bringing warm temperatures and possibly some rain to soften the snowmass enough that when colder temperatures return, the top layer of snow freezes rock hard.

By March the days are appreciably longer—deceptively longer, for March is also the time to beware of extra-heavy late snowstorms. As spring gains, significant changes take place in the winter snow cover caused by the erratic and slowly rising daytime temperatures. The snowmass becomes denser, more compact, and snow crystals become coarse and grainy. The result of these changes is often called *corn snow*, greatly beloved by spring skiers of both the downhill and cross-country breeds. The ideal corn snowpack is built on a firm base and will last as long as the nights remain cold, even though the days may have above-freezing temperatures. At night the corn snow turns to a rough-surfaced crust that will support great weight.

In the lengthening, warming days, a collection of melting forces work to reduce the snow cover. Most **snow-melt** takes place at the upper surface of the snow cover, where several forms of the sun's heat are at work. When the air over the snow surface is above freezing, and it is not windy, and the sun is masked by an overcast sky, only a very limited amount of melting can take place by direct conduction of heat from the warm air to the snow surface. Melting is increased considerably when the warm air is stirred up by winds so that convection aids in heat transfer. Air at the snow surface is constantly being replaced with warmer reinforcements through turbulent exchange.

Of course the great power behind all of these dramatic events is Old Sol himself, whose rays are falling more directly onto the snow now than they were in mid-winter. Immediate short-wave radiation from the sun is the main reason for snow melting. Newly fallen snow of mid-winter is intensely white and, as we have seen, it reflects about ninety percent of the sun's rays. Then, as the snow gets older and wetter, its reflective quality, or albedo, decreases to about forty percent. Dirt on the snow and other types of discoloration also make snow absorb the sun's radiation more easily, hastening the melting process. Powerful short-wave radiation can penetrate twelve to eighteen inches into the snow. Naturally, if the snowpack itself has become rather shallow the sun's rays may reach right to the dark-colored earth to add heat back into the budget and further increase the speed of melting.

Over much of the snowbelt of North America, winter comes to an end during the month of April. It's true that snowstorms can be expected all through the month and also in May, but usually the last of the real winter snow cover disappears in April. Warmer days and spring rains soon draw the frost from the ground, and the sun dries up the mud in unpaved country roads, usually by early May. Then winter has really ended—for a time.

2 Weatherproofing the Snow-Country House

IN SNOW COUNTRY, winter weather can take countless forms. Obviously, cold will concern you most, but snow and ice also threaten the effectiveness of the shelter your house provides—and you will want to do all you can to make your house as efficient and as snug as possible. You want to keep warm air inside the house and keep cold air out: it's that simple, and that complicated.

Immediately, you're interested in blocking the two main avenues of attack winter weather can use: (1.) Leaks in the exterior or Outer Skin of your house, allowing moisture and wind-blown debris (leaves and dust) to penetrate and damage its structure; and (2.) Leaks in the interior or Thermal Envelope that allow cold to enter and heat to escape, driving your energy costs up, disturbing your comfort, and perhaps causing damage to internal systems or furnishings. There is a good deal of work to be done to stop these leaks.

Who's going to do it? You can, of course, hire all improvements made by a contractor, but this chapter assumes—with a few noted exceptions—that you will do the work yourself. However, should you decide to have a contractor do a particular job, you will want to find a reliable one. Consult, first, the Yellow Pages to establish a basic list of contractors for the work you want done. To winnow the list down, take it to your banker (who will be interested in seeing you get your money's worth if he's giving you a loan for the job), the local chapter of the National Association of Home Builders or of the Associated General Contractors or (if there is one near you) the local office of any non-profit or government-funded home improvement assistance center. Once you have narrowed the list of contractors to three or four, ask each for a list of his past customers and find out from them how satisfied they were with the contractor's work. See how long each contractor has been in business—the longer the better, usually. Check with your local Better Business Bureau to see if complaints have been made against any of the contractors on your list; and get an estimate from each contractor for any work you think will cost over, say, $200.

What are you—or your contractor—going to do first to weatherproof your house? You're going to begin sealing breaks in the Outer Skin and Thermal Envelope, and you're going to start on the Weather Side.

The Weather Side of your house is the side facing the direction most of your bad winter weather comes from. In snow country, this usually means the north side; in actuality, you'll find that winter storms usually come most often from a sector of something over ninety degrees—perhaps from about northwest to about northeast. Your geographical location in the snowbelt, as suggested in Chapter 1, will have a lot to do with where your bad weather comes from, as will the topography of your particular location. If you live, for example, in Vermont on top of a hill, winter storms may come from anywhere in an arc from due west through north to due east, and the storms that leave the most snow may come from northwest or northeast. The north side of such a house is definitely the Weather Side as far as snow is concerned; its prevailing winds would probably come mostly from the west. (But most of its *wet* weather would come from the south and southwest.)

To find out where *your* weather comes from, observe. Watch the sky. Talk to your neighbors, especially the ones who have lived in your area a long time. Orient yourself and the location of your house in relation to the cardinal compass directions, and remember where they are on your personal horizon. Knowing where north, south, east and west are is essential in snow country, and it's useful mental baggage anywhere.

When you have found the Weather Side of your house, and you

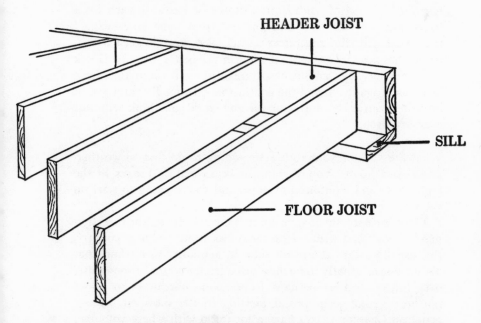

HEADER JOIST

SILL

FLOOR JOIST

Floor and sill construction in a modern house using relatively light structural members (above); and in an older house, using heavy timbers.

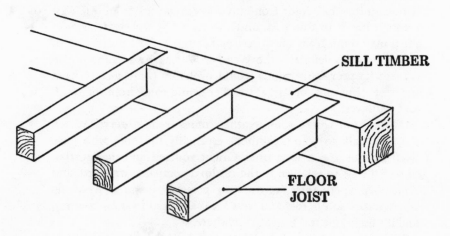

SILL TIMBER

FLOOR JOIST

know where your prevailing winds come from, you are ready to begin identifying and sealing the breaks in the Outer Skin and Thermal Envelope of your house.

THE OUTER SKIN AND THE THERMAL ENVELOPE

The Outer Skin of your house has several parts. From the ground up (the easiest order in which to work on them) they are: the foundation, sills and header joists (see drawing) ; the outside walls; all openings in the walls (windows, doors, vents, electrical and telephone service entrances) ; the roof and the openings in it (chimneys, vents). In good repair, the Outer Skin provides a physical barrier to wind, moisture, dust, insects, and (to a degree) cold.

Within the Outer Skin is the Thermal Envelope, the barrier surrounding the heated areas of your house that keeps heat and water vapor from escaping and cold from entering. From the top down (the most cost-effective order in which to work on them), its components are: the attic insulation (insulation between the attic floor joists, and in attic end walls, between rafters, knee studs, or collar beams) ; the insulation in the outside walls of all living spaces, including heated basements; the insulation in the flooring over an unheated cellar or crawl space; the insulation between any heated area and an unheated one (a storage room, garage, an overhang, dormer ceilings), including insulation in the walls of heated crawl spaces; and insulation around pipes and hot air ducting.

Together the Outer Skin and the Thermal Envelope keep your house weatherproof and snug and your family comfortable. They keep your energy costs down by preventing the escape of heat, and they keep your structural maintenance costs down by preventing the entry of dampness and cold that can damage the fabric of the building or cause freezing and rupture of water pipes. You, with the help of caulking, putty or other glazing compound, weather-stripping, paint, additional insulation and other materials, need to repair any breaks and remedy any inefficiencies in these barriers. Their condition, and the condition of your purse, will determine how much you'll need to do and how quickly you'll be able to do it.

REPAIRING THE OUTER SKIN

With winter coming on (it doesn't really matter what time of year it is: winter is always coming on), if the weather is reasonably warm and dry outside, you can start work on the Outer Skin (you can work inside to strengthen the Thermal Envelope when winter weather has set in).

In practice, your family's convenience, the age and condition of your house, your bank account, and common sense will tell you whether to begin inside or out. For the purposes of this chapter, assume you will start outside, on the Skin. We'll also assume you're starting on the Weather Side and working around the house as your time and circumstances permit.

If you have time or money to do only a few things initially, concentrate on the following for most immediate cost effectiveness: (1.) If your house needs major roof repairs, they should get first priority, for roof leaks can endanger the entire structure of a house; (2.) Put on temporary or permanent storm sashes; (3.) Caulk and weatherstrip all exterior windows and doors; and (4.) Bank your foundation. Do the last three in the order indicated. Little by little, work on sealing the other components of the Outer Skin until all weaknesses are eliminated.

THE FOUNDATION, SILLS AND HEADER JOISTS

Inspect your foundation inside and out for cracks and chinks through which light (and therefore air) can pass. Plug these with mortar or oakum if they are large, or stuff them with small tufts of fiberglass insulation material (wear gloves for this work; fiberglass can cause painful skin irritation). Check your work carefully with a candle or match flame to detect air movement.

The type and condition of your foundation will dictate your approach to the problem of insulating a cellar or crawl space. In particular, water in your cellar can render pointless any work you have done to insulate your foundation. Water that runs in and out of your cellar during periods of heavy rain will quickly destroy any insulation in its path. If your cellar is damp from ground water seepage or roof run-off, you may be able to alleviate the problem with eavestroughs and downspouts or other drainage measures (see The Roof for a discussion of eavestroughs). In any case, however, water problems may determine, in your situation, the degree to which a cellar or ground level foundation can be insulated.

Check your **cellar windows** and your outside cellar entrance (if any) carefully. They are holes in your foundation, and you should make them tight to prevent heat loss and cold air entry. Seal any cracks between casings and foundation with caulking or oakum, and replace any cracked panes; repair missing or cracked putty or other glazing material with new. Paint the window sashes and casings or at least coat them regularly with a wood preservative like pentachlorophenol products.

If your cellar is heated and relatively moist you may find that your cellar window sashes are slowly rotting from condensation. Heated air carries more moisture than cold air, and the moisture will condense out of the air onto cold windows. To alleviate this problem you can construct an inside storm sash, from wood and glass or rigid plastic for a permanent job, or from clear flexible polyethylene sheets for a temporary repair. In either case, your goal is to create a dead air space between the warm air in the cellar and the cold outer sash so that condensation is minimized. If you are using polyethylene sheets, cut them one inch larger than the window openings on all sides (two inches on all sides if you're installing sheets outside). Tape the sheet to the window frame with masking tape or broad, heavy-duty duct tape for an inside installation; for an outside installation, double the two-inch overlap back on itself and tack the sheet to the outside of the casing with a staple gun. Then nail wooden slats over the staples all around the sheet to prevent the wind from stripping your sash away.

BANK YOUR FOUNDATION

As a temporary measure (one you can renew each fall), bank your foundation. This is an old but effective way of cutting down on drafts through your foundation, and of using snow as insulation against the outside of your foundation. But to be effective— and to avoid moisture damage to your house—banking must be done properly. Use builder's sheathing paper, also called rosin paper. This has low vapor resistance (that is, it is *not* a vapor barrier), and comes in thirty-six-inch rolls. Attach one edge of a strip of sheathing paper to the base of the wood siding of your house, overlapping the siding at least six inches and extending to the ground and lying along the ground at least eighteen inches. Fasten the paper to the house siding using wood lath nailed at an angle to the horizontal, to increase vertical holding area and allow water run-off. Tape all vertical seams with duct tape.

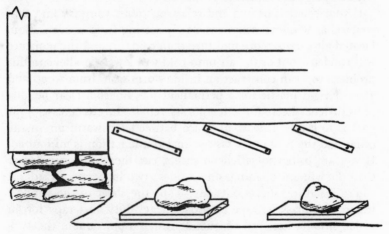

Placement and securing of foundation banking.

Weight the bottom outer edge of the paper with boards, further held down by rocks or bricks, to keep it tight to the ground and prevent the wind from getting under it, lifting it, and tearing it away. As the winter wears on, the paper, which has little wet strength, will gradually get soggy during thaws. Rocks or bricks alone as weights might puncture it—use boards to spread the weight and holding area. Watch out for old nails. If necessary, use two overlapped widths of sheathing paper for adequate coverage of siding and ground. Lap the upper layer over the lower one.

The primary purpose of the sheathing paper is to act as a wind barrier to drafts that might penetrate your foundation. However, assuming that your foundation is porous enough to require such a barrier, it will also release warm air from your cellar to the outside. If the moisture in that air cannot penetrate the wind barrier and escape, most of it will condense inside the barrier on the outside of your foundation and on the base of your siding, hastening wood rot and causing damage to any mortar on the outside of the foundation. This is why experienced householders use permeable paper banking rather than polyethylene sheets or tar paper; the latter are vapor barriers and will trap moisture. The proper place for a vapor barrier is *inside* your cellar, against the foundation wall, to prevent warm air (and its water vapor) from leaving the cellar. Once it has left the cellar, through leaks in your foundation, you want it to keep right on going to make sure its moisture isn't left behind to cause problems.

With this wind barrier in place, you can allow (or encourage) snow to drift and bank solidly against the foundation, further enhancing the barrier's effectiveness and providing additional insulation. Some householders pile hay, straw or fallen leaves against the foundation to provide extra insulation, and lay the paper over them. In contrast, many old-timers laid up hay or pine boughs *on top of* the paper to trap snow and hasten its build-up (if you try this, watch out for sharp twigs that could puncture the paper). Hay or leaves, which also work as a snow trap, will turn soggy with spring and perhaps speed up the disintegration of the sheathing paper (which you'll have to replace each season anyway). And if your barrier disappears too soon, there will come an April day when the winter you thought was over sneaks up again, and you'll wish that banking (rustic and unsightly, perhaps) were still effective.

As a practical matter, if you're going to work on the outside walls, doors and windows, you'll do that first, and save the banking for last, to avoid damage to the paper wind barrier.

THE OUTSIDE WALLS

You must make sure that your house siding presents an unbroken surface to the weather. Caulk under loose clapboards and nail them down. Spot clapboard nails that have worked up and drive them down again. Caulk cracks in the siding between the siding material and corner trim and door and window casings. If you have an outside chimney, inspect the crack between it and the house siding, and caulk it tightly. Check the mortar and masonry for cracks that could admit water, and seal them to prevent freeze damage to the chimney. As you go, check painted surfaces for signs of peeling, mildew, dampness—these can be evidence of internal wall moisture that may indicate leaks in your Thermal Envelope.

Doors, windows, vents and other openings in your walls must be tight. Wall insulation can do only so much: a good deal of heat loss from well insulated houses can take place at doors and windows. If possible, select one door as your winter entrance and seal the others off until spring. This will minimize heat loss when the door is opened, and will cut down on the required clean-up of tracked-in snow and mud. The chosen door should be located, ideally, on a side of your house protected from wind and from snow slides from the roof. Although your other doors are sealed, don't block them, and make sure they can be opened. You want more than one way out if you have a fire.

Check all **doors** to the outside to be sure they fit their openings properly and do not bind. Straighten a binding door in its casing by tightening hinge screws, shimming hinges, or (as a last resort) planing the door, to ensure a square fit. Install weatherstripping to seal the sides and bottom of the door against drafts.

Weatherstripping comes in a variety of forms and materials. Simplest are felt or adhesive-backed foam strips which are tacked or stuck to—for example—door frames where the closed door comes in contact with the frame. There are also metal, rubber, and vinyl products, in strips, tubes and grooves, and metal-rubber and wood-foam combination weatherstripping in which the rubber or foam is bonded to a metal or wood backing which is nailed down into the area to be sealed. In each case, when the weatherstripped door or window is shut, the weatherstripping material is compressed, sealing the opening between the door or window and its frame. The simpler felt and foam strips are cheaper than other forms of weatherstripping, and far easier to install; but the more expensive types are more effective and longer lasting.

If possible, as you go about tightening your doors—especially on the weather or prevailing wind side of your house—install a storm door (which you will weatherstrip also) to provide a layer of dead air that will better insulate your main door against cold. Hinge it, if possible, on the side opposite that of the existing door hinges, to minimize the opening required to go in and out. Better yet is a storm house, a relatively extensive project which I will discuss under Finishing Touches.

Inspect your **window sashes** carefully for cracked panes and cracked or missing putty, and repair or replace them as necessary. Caulk cracks in the window casing, between the casing and the siding of the wall.

Caulking compounds and sealants presently on the market show even more variety than types of weatherstripping. There are at least ten different base materials used in caulking compounds, including latex, neoprene, polyethylene, silicone and polyurethane. Some caulking compounds must be applied in warm weather; some can be applied any time. Some are for use on protected surfaces only. Some adhere to virtually all materials; some only to concrete or wood. Some can only be successfully applied to primed surfaces, some can be put on anywhere. Some will last for twenty years, some for only one or two. Caulking compounds also vary widely in cost, and it is worth your while to consult a hardware or paint dealer on what caulking compound is best suited to any job you have in hand (see also the list of books at the end of this chapter).

When you have finished caulking any cracks around your windows, check for small drafts inside, as you did around your house's foundation. The tiny air leaks you miss now during warm weather, when driven by a twenty-mile-per-hour wind at twenty degrees Fahrenheit will feel like needles.

Weatherstrip all window sashes to eliminate drafts at their sides and where a top and bottom sash join. Use spring-metal stripping on double-hung windows whenever possible—it's the most effective, and invisible when properly installed. Other kinds of stripping are easier to install, but will be visible and won't last as long.

Install **storm sashes.** As on your cellar windows, you can do a fast temporary job with polyethylene sheets and wood lath, or you can do a permanent job with glass or rigid plastic and wood or aluminum frames you build yourself. Storm windows are expensive; building them yourself will make a major savings. It's time-consuming, however, and often difficult to frame and hang a homemade window accurately; and a storm window that doesn't fit is scarcely worth the trouble it took to build it.

You may well be better off buying triple-track aluminum storm-sash combinations (including windows and screens in the same unit) for ease and speed of installation—but discuss carefully with your supplier the measurements required if you plan to do the installation work yourself. Double-track (windows only) storm sashes are cheaper. Be sure to inspect the available styles carefully for good quality. Pay particular attention to corner joints: they should be air-tight, and overlapped joints are usually better in this regard than unwelded mitered joints. Be sure the sash tracks are deep enough and the weatherstripping adequate to minimize air leakage around the sashes; compare several varieties before ordering. Inspect the locks and catches for general quality—they have a direct effect on durability, and provide a good indication of the overall quality of the window.

Permanently installed aluminum storm windows are effective

and require little maintenance or labor (except for cleaning) because they are stored in place: the storm sash simply slides up to the top half of the window when not in use. For most effectiveness, however, you might want to consider old-fashioned, **wooden storm windows.** Less expensive than aluminum, they require maintenance (paint and glazing compound), and a storage place during the summer. Also, it can be a job to put them on and take them off. Properly hung and weatherstripped, however, they surpass aluminum storm windows in resistance to heat loss (metal being a better heat conductor than wood). In addition, if wood storm windows are adequately vented, with two or three one-inch holes bored through the bottom of the frame (closed by a pivoting or sliding panel), vapor can escape and condensation largely be avoided—provided, of course, your inner sash is well glazed and weatherstripped. If yours is an old house with a full set of wooden storm windows, or if you can acquire a discarded set that can be made to fit (they should fit tightly *inside* the casing), count yourself lucky. Unless you're prevented by health or strength from coping with the twice-yearly chore of mounting and dismounting wooden windows, you may want to reconsider the notion of substituting aluminum ones. Consult your budget and compare the relative costs.

Now, in theory, your outside walls and doors and windows have been made impervious to wind and water. What about vents for your clothes dryer, kitchen fan, or air-conditioner? The place where each of these comes through the wall is a hole in your house's Outer Skin, as is the hole for your electrical service entrance, telephone wire, outside television antenna wire, outside water faucets, oil tank filler or vent pipe, and similar installations. Make sure any gaps around these wall penetrators are sealed to prevent drafts from entering.

THE ROOF

Leaks through which moisture can penetrate will create more roof problems than air leaks, usually. Leaks can develop in the fabric of the roof, in the flashing around openings in the roof (for your chimney, a vent pipe, an air vent, or a skylight) ; or in the flashing at corners or in a valley (the angle where roofs that slope in different directions meet) .

Finding leaks in any of these roof locations may be difficult, because the water may follow a twisting course through all the layers in a roof (modern roofs usually have more layers than old ones) and emerge into the house in a spot far from its source. Difficult or not, it can be done. Look for pinholes of light through the roof, in flashed areas or between the sheathing boards in old roofs. Mark them for outside repair by poking a small wire through. Check the roof from outside, for curled, broken or missing shingles or slates. Look for roofing nails which have worked up (especially on metal roofs) and drive them down again. On a flat roof, check for blisters, tears or cracks in the roofing material. If you have a flat-roofed house in snow country, your problems are compounded by the fact that you will probably have to shovel the roof fairly frequently during the winter to prevent snow build-up that could collapse the roof. Shoveling a flat roof usually causes leaks or weakens the roofing surface slightly—eventualities that will have to be dealt with when the weather permits.

Check your roof, especially if your house is old, to be sure that the normally flashed areas mentioned above really *do* have flashing. Check the roofing cement around the edges of flashing for signs of cracking and crumbling, and renew the cement where necessary. Check walls and ceilings inside for discoloration and stains, and roof rafters for the white stains of mildew that can indicate the presence of dampness.

Check the flashing in gutters or at the eaves for rusty stains. These indicate corrosion of the flashing (especially if it's galvanized steel) , which should be removed with a wire brush. Protect the flashing with metal primer and aluminum paint. *Galvanic corrosion* results when dissimilar metals meet. Make sure that flashing and nails are of the same metal, and—if you make a repair—that new metal matches old.

During the winter, check the edges of a sloping roof for **ice dams.** These are created when snowmelt, caused by warm air seeping through the upper reaches of your roof, re-freezes. The presence of warm air under your roof may be the result of a leak in the Thermal Envelope of your house, usually in the attic insulation. The snowmelt runs down the roof to the colder edges, where it re-freezes.

The existence of an ice dam is a signal to you to do four things quickly: (1.) Remove the ice dam, taking care not to cause further roof leaks in the process; (2.) Strengthen the Thermal Envelope—probably by increasing attic insulation; (3.) Place pans or buckets to catch the drips already caused by the ice dam; and (4.) Notice leak locations so you can make repairs in warm weather. You may find it necessary, in addition, to add flashing to the eaves so that snow and ice won't stick, thus preventing ice dams from forming.

The best time to remove snow that has collected on a roof that is subject to the formation of ice dams is within a half day of its falling—after all the snow that is going to remove itself from the roof by sliding has fallen off, and before what remains has had a chance, by partial melting, to become heavy, sticky and icy. Use a long-handled scraper, rake, hoe, shovel or brush to clear the roof. Don't get onto the roof unless it is absolutely necessary: work from the ground or from a ladder.

A temporary solution to roof ice dams may be **electric roof-melting cables.** When the roof is clear, lengths of cable are clipped onto it in a zig-zag in places subject to icing up. When ice and snow begin to accumulate, you turn on the cables, which heat up and melt off the accumulation.

While you're up there on the roof, assuming that you've located all leaks and that the roof is now tight, check on the condition of your **chimney.** Are the flashings tight, without holes, firmly anchored? Is the mortar cracked or missing, especially in the top course of bricks? If it is, water can seep down into the cracks and freeze, gradually splitting the chimney. Have an experienced contractor point up the chimney (that is, repair the mortar) and build a mortar or masonry cap to keep water out of the chimney and from penetrating the top course.

How is your television antenna anchored? If it's attached to your chimney and you live in a windy location, it may jerk and whip in the wind enough to loosen bricks and mortar. Consider moving it to a gable end where it can be attached firmly with a bracket.

Eavestroughs or gutters are a subject of controversy among snow-country householders. Proponents say that, properly installed, gutters can direct snowmelt away from house foundations, while others claim gutters, by holding frozen water, allow water to back up on the roof where it can freeze and damage roofing. Some houses are fitted with brackets under the eaves so gutters can be hung up in the spring and taken down again before winter comes. If your house has permanent gutters, and you notice that they trap frozen snowmelt from your roof, get rid of them—they will create more problems than they will solve.

If your gutters are not a source of trouble, and you want to keep them, check them each year before winter comes. Make sure they are clear of debris and that the downspouts are not clogged. Sand rust spots to bare metal and coat with metal primer. Be sure that the gutters are in alignment, and check to be sure they drain properly by flushing them with a hose. Gutters can be knocked out of alignment, or torn off the house, by snow sliding off the roof. To avoid this, gutters should be installed low, with the front edge of the gutter well back below the slope line of the eaves, so that sliding snow will clear it. Once your gutters are installed so sliding snow can't tear them away or twist them, be sure that the downspouts do not discharge directly onto the soil near the foundation. Lead the water away from the foundation by means of a splashblock, a perforated plastic hose, a buried terra cotta or plastic pipe ("tile") or even a spare length of gutter.

In making roof repairs of any kind, use extreme caution. Don't attempt any roof work in cold, windy or wet weather. Cold makes roofing materials brittle, and dampness makes them slippery. Wear non-slip shoes or sneakers and clothing that allows you to move easily. To reach the roof, use a ladder long enough to reach above the eaves, so you won't have to step over the top of the ladder onto the roof. Get someone to hold this access ladder as you climb. Never lean out over the side of the ladder to work on the roof. Keep your hips between the ladder's rails. Have your helper hand you tools and materials once you are up on the roof. Use a "chicken ladder" (a twelve-inch board with one-by-two-inch horizontal wood cleats) on brittle roofing materials like slate, asbestos or tile. On a steep roof, use a roofing ladder which hooks over the roof ridge; such a ladder will spread your weight on the roofing material and give you secure handholds. Finally, if you are afraid of heights or your roof is unusually steep, leave the job to an insured contractor.

FINISHING TOUCHES TO THE OUTER SKIN

When you have made the Outer Skin largely impervious to wind and water, you may wish to consider a few finishing touches.

Add a **storm house** around the outer door you've chosen for winter use. Its purpose is to act as a sort of air lock, to prevent cold outside air and precipitation from blowing directly into the heated interior of your house, and to prevent the escape of warm inside air. It's an exterior structure, built in sections for seasonal removal, that fits tightly to the outside walls of the house around the casing of the door (see drawing). The storm house should be caulked and weatherstripped each season between sections and where it joins the main house wall. Don't make it too small. Leave at least room enough so that a person going out can open and close the outer house door *before* opening the exterior, storm house door. If possible, build the exterior door into a sheltered side of the storm house so that a wind blowing directly toward your outer house door doesn't have a straight path of entry.

A well made storm house can be a relatively expensive project if you have a contractor build it, less so if you can do it yourself. You may, however, find your house suitable for a cheaper solution, especially if it's an old one, in which case it may have been built to include an interior storm vestibule or "mud room." If possible, close the doors to all rooms or passages leading in from

Removable storm house.

the entrance door. Install a storm door on the outer door. Provide coat hooks and mats for foot-wiping near the outer door in the mud room. The drafts that do penetrate the house when the outer door is opened will be confined largely to the mud room, and of course the warm air in the interior rooms will be kept in.

Where is your outer entrance door located? Ideally, on a sheltered side of your house, perhaps even the south side—but unless that sheltered side is also a gable end, your entrance door is subject to blockage by snow slides from the roof. In some instances, the amount of snow liable to slide off the roof isn't great enough to be a problem. The overhang of your roof should pitch most of the fall beyond the doorstep. But after a major storm, one quick, rumbling snow slide can be enough to require a lot of hard digging, and could block your regular exit door until someone is able to dig it out—a definite problem for elderly persons. How to avoid it?

Have your roofing contractor fashion the flashing at the eaves so that a ridge of flashing in the form of an inverted V rises from the normal slope of the roof over the entrance door (see drawing). The ridge need not be more than about three or four inches high, nor the width of the V at its open end greater than the door opening, as its primary purpose is to divert most of the sliding snow to either side of your doorstep. If a little snow overrides the ridge and falls on the doorstep, it can be removed easily by shoveling or sweeping.

If your entrance door is not in a gable end you should ask yourself whether a door can feasibly be added to a gable end without exposing the door to the prevailing winter wind. Relatively major construction is required, best handled by a reliable contractor, and you should think carefully about this alternative, choosing it only if your area is subject to snowfalls of sufficient magnitude to block your entrance regularly with roof slides.

Roof flashing snow deflector in position over door.

REPAIRING THE THERMAL
ENVELOPE: INSULATION

Improving the efficiency of your house's Outer Skin is primarily a matter of plugging holes; improving the interior Thermal Envelope is primarily a matter of insulating. The purpose of insulation is to trap air and hold it motionless ("dead air") in minute spaces so it will not convey heat out of your house.

Insulation may be "loose fill," fiber, board or foam. Loose fill insulation—it can be of any loose material, like sawdust, perlite or vermiculite—is easy to install: you simply pour it into the space to be insulated. It is the least effective insulating material, however. Fiber accounts for ninety percent of the insulation sold in the United States—generally in the form of fiberglass blankets or batts, or so-called "rock wool," a mineral product. Fiber insulation is easy to install and effective. Board insulation material comes in rigid synthetic panels, one-half to one inch thick. Its main drawback is that it is flammable; the panels should always be covered with a fire-resistant sheathing like gypsum wallboard. Foam is the most effective insulation material and also the most expensive, since it is usually installed by professional contractors using specialized machinery. It is squirted into walls, floors and ceilings through holes, and allowed to set in place.

What kind of insulation, and how much of it, you need, depend on a variety of factors, including your location, the age of your house and whether or not it has any insulation in the first place, your heating bills, your bank balance, availability of various insulation materials in your area, and whether or not you have home air conditioning. For houses in the northern United States and adjacent Canada, most authorities recommend six to eight inches of fiberglass insulation in the ceiling below an unheated roof or attic, four inches in outside walls and six inches below floors over unheated cellars or crawl spaces. Several of the books listed at the end of this chapter will help you evaluate more precisely your own insulation needs; you should also consult building and insulating contractors, and home heating suppliers, in your area.

In installing any insulation it is essential to provide a **vapor barrier.** Whenever the temperature inside your house rises above the outside temperature, warm air will move out of your house. Water vapor in that warm air will condense when the air is cooled on passing through your outside walls. The resulting dampness will reduce the effectiveness of the insulation, promote rot in the structure of your walls, and cause inside paint and wallpaper to deteriorate. The purpose of the vapor barrier is to prevent water vapor from passing into the insulating material behind it.

Vapor barriers may be built into some insulating materials (the popular fiberglass blankets are backed up with metal foil or heavy paper for this purpose). For other kinds of insulation (especially loose fills, which absorb moisture easily), a sheet of polyethylene, covering the entire insulated area, is tacked over the insulating material for a vapor barrier. Certain foam insulating materials are claimed to provide their own vapor barrier, no additional materials needed. However, some builders feel that the adhesion of foam insulation to the wall structure may be poor enough to allow vapor to seep between the insulation and the structure, necessitating a separate vapor barrier.

Vapor barriers are installed on the warm side of the insulating material they are to protect. Therefore, if you are installing insulation from the inside of a wall, the vapor barrier goes on last, and faces the heated living space. If you are installing insulation from the outside (for instance, under a floor from an unheated cellar) the vapor barrier goes on first—before the insulating material—again facing the heated living space.

WHERE TO INSULATE

If your house is uninsulated, or if you decide its insulation is inadequate, you will need to add insulation to all partitions that separate heated spaces from unheated spaces, open or enclosed:

attic, outside walls, floors, walls of garages, porches or similar unheated areas—all require insulation. Because of the tendency of
warm air to rise, "cap insulation" in the ceiling over the upper
rooms of your house is most effective in reducing heat loss and
fuel bills. Keeping heat from escaping through the ceiling into an
unheated attic, if any, and through the roof will also greatly
lessen the chance of snowmelt and the resulting formation of ice
dams on the roof outside.

Determining whether **attic insulation** is needed is fairly easy
in houses with attics. If your attic is unfinished, it is easy enough
to check the amount of insulation you have simply by inspecting
the area between the attic floor joists. If the attic has a butted
board floor, carefully pry up boards to get a clear idea of how
much insulation is beneath them. If the floor is tongue-in-groove
boards, don't try to pry up a board (if you try and find it difficult
to pry up a floorboard, it is likely the floor is tongue-in-groove) ;
rather, drill a one-half-inch hole in the middle of a board (taking
care to miss the nearest floor joist) and probe gently with a wire
or pencil, first for the top of any existing insulation, then deeper,
for the bottom, to find the depth of the insulation.

The condition of your attic, the amount of existing insulation
you find, and your ultimate plans for the attic space will guide
you in deciding how to proceed in insulating an unfinished attic. If
the attic is presently unfloored, or if you can remove the flooring
easily, and you don't intend to turn the attic into a heated living
space in the future, the easiest and most effective method of insulating is to insulate the attic floor—for instance by adding fiberglass blankets (vapor barrier backing *down,* or facing away from
you) to the area between attic floor joists. If you can't raise the
attic flooring you can insulate the roof with fiberglass blankets
hung between the rafters (vapor barrier backing *up,* that is,
facing toward you) , and between the wall studs in the gable ends.
The easiest way to insulate your roof is to install insulation from
the peak of the roof down to the top plate of the wall where roof
and wall meet. More effective, however, is to nail up collar beams
(horizontal ceiling joists between rafters) and bring the insulation up from the attic floor to the level of the collar beams, then

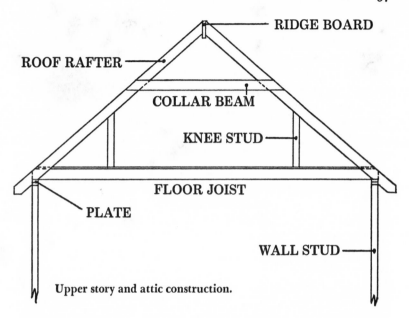

RIDGE BOARD

ROOF RAFTER

COLLAR BEAM

KNEE STUD

FLOOR JOIST

PLATE

WALL STUD

Upper story and attic construction.

hanging it from the collar beams in the space between them. This leaves an uninsulated area just under the peak of the roof, which, vented, lets hot air escape in summer and also serves to vent water vapor. If you intend some day to finish your attic and include it in the heated part of the house, you will want to add knee studs at right angles to the collar beams and attached to the attic floor (see drawing). You then install the insulation between the knee studs and collar beams, and, in the attic floor, only in the area between the roof and the knee studs.

Once you are satisfied that your attic is adequately insulated, you can turn your attention to your outside walls and floors. These extend from the roof of your house to the basement floor.

In new or unfinished houses, **insulating outside walls** is a relatively easy matter: fiberglass batts or blankets are hung in the open studwork from the inside. In an older, finished house, unless you want to bear the expense and trouble of removing the inside wall covering to expose studs or posts, the best method of insulation is to blow material into the space between the outer and inner wall coverings from the outside. If your house has walls of masonry or solid wood, you will have to build so-called "furring strips"—thin wood strips fixed to masonry walls with adhesive or masonry nails—on the inside walls and then nail board insulation onto them.

Loose fill insulation can often be blown into outside walls by the householder, using rented equipment; and even when the work has to be done by a contractor, you can accomplish much of the preparatory work yourself, for instance the boring of outside holes in walls to admit foam insulation. Whatever work you are able to do yourself will save you money on your contractor's time, but consult closely with him from the start.

Vapor barriers for outside wall insulation always face inside the house. When insulation is installed from inside the house providing a vapor barrier is easy, but when loose fill or foam is blown in from outside, special provisions for the vapor barrier are needed (unless you decide to rely on the self-protecting properties of foam insulation). In these cases you must create a vapor barrier by painting all inside wall surfaces with low permeability paints (consult a paint supplier or painting contractor) and making certain all interior cracks around door frames, windows, electrical outlets, etc., are tightly caulked.

In a house with an unheated—or poorly heated—space under the heated area (crawl space, cellar, ground floor garage) insulation should be installed beneath the floor of the heated area. Make sure the foundations of the house have been banked and otherwise made tight outside. Attach fiberglass blankets or batts between floor joists in a cellar, garage or high crawl space. In these installations, the vapor barrier goes *up*, facing away from you and toward the warmed rooms above. Hold insulation in place with wire braces or chicken wire. If your crawl space is too small to work in, you may have to resort to foam insulation. In any case, lay sheets of polyethylene on the dirt floor of the crawl space, overlapping the sheets six inches. Secure the ends of the sheets to the foundation wall with duct tape.

REMOVING EXCESS MOISTURE

As you gradually tighten your house against drafts, cold and mois-ture from the outside, you may discover that along with a reduced fuel bill you have let yourself in for a problem: the accumulation of excess moisture inside your house, which can no longer escape via its customary wasteful routes through cracks and crannies. Especially if your house is small, excess moisture can lead to an unpleasant damp sensation, and it can cause mildew and peeling paint and plaster. To combat it, you may need to add vents and fans to your house as you tighten its Outer Skin and insulate it.

Since warm air can hold more moisture than cold air, and warm air rises, the most effective place to vent warm air from your house is in the roof area. Gable end vents under the roof peak, vents cut through the roof, vents built into the roof ridge, and vents located under the eaves, are all possibilities. All can serve to vent your house's attic, which it is especially important to do. Attic vents should be open all year around. In summer, they cool the house. In winter they prevent condensation from forming when air passes from the warm lower floors of a house into the colder attic.

In some houses, especially in rooms subject to concentrations of water vapor (kitchens, bathrooms), exhaust fans may be needed to remove moisture (although the old expedient of simply open-ing the top window sash and the storm window for a few minutes still works wonders in venting moisture). Exhaust fans should be vented to the outside. It is best not to follow the common practice of wiring fans to a room's light switch so that the fan will turn on whenever the light is turned on. After all, fans blow heat out of the room along with unwanted moisture. Give the fan its own switch, and *you* decide when it runs.

OTHER HEAT-SAVING STEPS

In addition to caring for the Outer Skin and Thermal Envelope of your house, there are other steps you can take to save heat when winter comes.

Make certain you have **heavy curtains or drapes** for all windows, and draw them at night. Radiant heat can escape from your house through window glass at night, creating a significant heat drain. Shades and drapes are effective in counteracting this loss. Windows on shaded sides of your house can also be curtained during the day to conserve heat further.

Another important route by which heat can escape from a house is an **open fireplace** not in use. After your evening fire has died down the fireplace and chimney become a large cold hole in your house, through which heat gets out easily. Keep your fireplace's damper closed when there is no fire, and get a metal or asbestos plate to fit over the fireplace opening when you leave the fire and let it go out. The plate will prevent heat from escaping via the cold fireplace after the fire dies.

Floor or wall registers for warm air ducts or cold air return can be sites for heat loss in houses with forced warm air heat. Make sure the registers fit tightly in their receptacles, and that their edges are sealed.

In addition to insulating the exposed areas of a house, most northern householders should **insulate hot water pipes** and warm air ductwork in unheated areas like cellars. Insulating pipes helps prevent freeze-ups and reduces heat loss, cutting hot-water heat expenses. Insulating ducts reduces radiant heat loss from the ducts, allows more heat to reach the living space, and cuts heating expenses. Pipes can be insulated by wrapping newspaper around them and taping it in place. More effective are the cylindrical pipe sleeves with canvas backings, slit vertically, which fit easily around pipes. To **insulate hot air ducts,** use specially-made thin duct-blankets, or wrap pieces of fiberglass room insulation around them horizontally (vapor barrier out, facing you), taping seams with duct tape.

A time-honored way of preparing snow-country houses for winter is to **close off little-used rooms.** This has the effect of reducing the size of the area to be heated, thereby reducing the amount of heat required. Consider whether there are seldom-used rooms in your house which could be kept shut off all winter long. If there are, close off the heating ducts or pipes that lead to those rooms. Close them off as near the furnace as possible, with valves or dampers provided for the purpose: you want to avoid needlessly heating the duct or pipe that leads to an unused room. Seal the doors leading from the cold room to heated rooms with masking tape or with a temporary caulking rope like Mortite, which can be neatly removed the next spring when the room will be opened up again. If water pipes pass through the cold room, make sure they are well insulated to prevent freezing.

Outside the cold room, place a cloth cylinder filled with sand against cracks under the door which are too large to caulk. Pack the sand into the cylinder loosely enough so that you can mold the cylinder to fit into corners and crannies under the door: if it is packed too tightly it will just lie there like a stick.

AROUND THE HOUSE

Anyone whose house requires improvement in efficiency of its Outer Skin and Thermal Envelope should attend to these improvements before he turns his attention to the area around the house. No changes in your yards or other outlying areas will affect your winter comfort as noticeably as will insulating and tightening up your house, inside and out. Nevertheless, there are projects you can carry out at a distance from the house which can make the house itself more winter-worthy by sheltering it from wind and snow.

Shelterbelt plantings or windbreaks on the Weather Side of your house can slow down winter winds and make it easier to keep the house warm. Some authorities claim that a good shelterbelt between your house and the prevailing winter wind may reduce fuel consumption by as much as fifteen percent.

The ideal density for a shelterbelt planting is fifty percent—that is, half the total area of the vertical screen provided by the trees and shrubs in the belt should be solid, half open. Which plant species are used to achieve such a screen vary from region to region. Consult local nurserymen or county foresters for suggestions on species suited to your area.

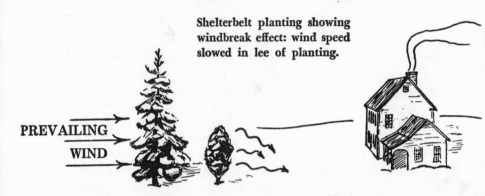

Shelterbelt planting showing windbreak effect: wind speed slowed in lee of planting.

PREVAILING WIND

The size of the area protected by a shelterbelt is proportional to the height of the plants in the belt. The most effectively protected zone will be the area in the lee of the plants extending for a distance equal to four or five times the plants' height. In this zone a twenty-five-mile-per-hour wind outside the shelterbelt will be slowed in velocity to about seven miles per hour in the lee of the windbreak.

The placement of a shelterbelt planting to protect your house will be dictated by the height attained by the trees and shrubs in the planting. Taller plants will be further from the house. Usually a well designed shelterbelt consists of two or three rows of plants of different species, the taller trees being laid out to windward of the lower shrubs.

Obviously, the benefits of shelterbelt planting are mostly long-term. Shelterbelts you plant now won't reach their maximum effectiveness until the trees and shrubs in them mature, though they will in most cases do some good from the outset.

Although they will not offer much protection to your house from wind—as shelterbelt planting will—snow fences correctly placed can save you much time and effort spent in domestic snow removal. Placed at right angles to the prevailing winter winds, snow fences slow those winds, causing them to drop snow in the lee of the fence (shelterbelt plantings have the same effect). Most snow fences are made of light vertical wooden slats strung together on horizontal wires. As with shelterbelts, snow fences should have a fifty-fifty ratio of solid to open space for maximum effectiveness. Fences are stretched, staked and guyed in place at a distance from the area to be protected. The most important point in setting up the fence, besides its correct orientation to the prevailing winds, is that sufficient space be left between the fence and the area to be protected (yard, driveway, road) ; for snow will accumulate in the lee of the fence, on the protected side. As a general rule, allow eleven to sixteen times the height of the fence for a collection zone.

Snow and ice removal around the house are a continuing activity for all snow-country householders, even those with shelterbelts or snow fences. There are two chief ways of dealing with large expanses of snow and ice: you can move them or you can melt them.

If you move snow out of your walks, driveway, etc., either by shoveling, or with a snow blower or plow, remember that snow falls all winter long. Begin by leaving the snow you move at an ample distance from the area you want to keep clear; this will insure you have somewhere to put snow left by future storms.

To melt snow and ice on your driveway or walks there are several commercial products available, including rock salt, calcium chloride and urea. Calcium chloride pellets work faster than salt or urea to melt ice, but they must be stored in a tight bag or box, for they tend to absorb moisture. All three melters can damage concrete, kill grass under the melted snow, and irritate the unprotected feet of your household pets. A more benign—and cheaper—way to deal with ice on walks and drives is to scatter sand, wood ashes, or sawdust over it. These will make the ice less slippery, and, because they act to darken the snow and ice, reducing its *albedo* (see Chapter 1), they cause it to melt faster. Whatever snow-melter you use, you will need effective doormats and boot scrapers at your entrances to avoid having the stuff tracked into the house.

If you would avoid the work of shoveling paths and walkways about your house, and you own a pair of snowshoes, you can often dispose of small and moderate accumulations of snow by packing them down in the paths. Walk the paths over a couple of times in your snowshoes. Soon they will be as passable as if you had shoveled them out—all at much less expense of time and muscle.

GOOD READING ON WEATHERPROOFING

Petersen, Stephen R. *Retrofitting Existing Housing for Energy Conservation: An Economic Analysis.* 1974. Published by the National Bureau of Standards, U.S. Department of Commerce. Available from the U.S. Government Printing Office for $1.35 (No. C13.29:2/64).

U.S. Department of Housing and Urban Development. *In the Bank . . . Or Up the Chimney? A Dollars and Cents Guide to Energy-Saving Home Improvements.* 1975. Available from the U.S. Government Printing Office for $1.70 (No. HH 1.6/3:EN 2/3). *Excellent on figuring costs and savings.*

Time-Life Books. Weatherproofing. 1977. Available at bookstores or from the publisher. *Excellent on materials available; good drawings show installation techniques clearly.*

3 Heating the Snow-Country House

HAND IN HAND with the measures discussed in the preceding chapter for keeping cold out and holding warmth in goes the matter of how to provide adequate heat, and this problem must be faced and dealt with as early as possible before winter sets in.

Although the meaning of "adequate heat" differs for different people, heating experts think in terms of what they call a **Comfort Zone**—a healthful household temperature maintained evenly throughout the day. This ideal temperature used to be considered 72 degrees Fahrenheit (22.2 degrees Celsius), though recent research has established that 65 degrees Fahrenheit (18.3 degrees Celsius) is, when coupled with proper humidity, pleasantly warm for the average healthy, casually active adult. If 65 degrees seems too Spartan to you, split the difference. A great many householders say they feel more comfortable physically since they began lowering the thermostats of their automatic central heating systems to 68 degrees Fahrenheit (20 degrees Celsius) at the onset of the energy crisis in the early 1970's. Even this small decrease in thermostat setting from 72 to 68 degrees Fahrenheit can, over time, result in appreciable savings of heating fuel: it has been estimated that a one-degree drop in thermostat setting can effect a three-percent fuel saving.

The body heat regulators of some members of your household may not function well enough to compensate for lowered heat during the day. Old and sedentary people, people with cardiac or circulatory problems, and infants probably will need more heat. They can take advantage of warm areas near sunny south windows on the side of the house sheltered from wind. In parts of the house frequented by those who need extra warmth, check care-

45

fully for drafts from cooler rooms or along the floor, and increase the heat if necessary.

Thermostats turned down at night will also reduce fuel consumption. A workable system is to set the thermostat five degrees lower at bedtime. A greater reduction, especially with a high wind-chill factor present, might demand an extra surge of heat the next morning to raise the temperature to the daytime norm.

IMPORTANCE OF HUMIDITY

Dry air must be heated more than moist air in order to be comfortably warm. "It's not the heat, it's the humidity," we say during a muggy heat wave, meaning that a high temperature reading would be more bearable if the air was drier. In cold weather we apply this principle in reverse, by adding moisture to the air in our houses to provide a comfort zone which is actually several thermometer degrees below what it would need to be if the air was dry. The ideal humidity for a temperature of 65 degrees Fahrenheit is forty percent; for 68 degrees the humidity can be slightly lower.

A simple hygrometer or other humidity gauge need not be expensive. One placed in each of your most heavily used rooms will soon repay the total cost in fuel saved: when you see that the relative humidity has fallen below the optimum, you can add water vapor to the air—and turn down the thermostat. (Keeping track of humidity, and making sure that it is adequate, will also benefit the respiratory passages of everyone in your house.)

Boosting humidity is usually necessary in houses with forced warm air central heating systems, and in houses which use wood- or coal-burning space heaters. For increasing humidity, automatic humidifiers can be installed on most central warm air heating systems. There are also a number of portable electric humidifiers available from mail-order houses or from heating and electrical appliance dealers (costs range from about $35 to $250). Long shallow metal pans can sometimes be placed along the tops of electric heat baseboard radiators: kept full of water they make excellent humidifiers, though the pans may have to be specially made.

Other means of increasing humidity, room by room, include putting open pans of water on top of radiators or space heaters, and providing fresh water vapor with a trigger spray bottle several times a day. Potted plants by themselves seldom add adequate humidity to air; set plants in long shallow trays full of pebbles and keep the pebbles in about an inch of water. A rack of freshly laundered clothing or linen air-drying in a corner will also add humidity to room air.

CENTRAL HEATING SYSTEMS AND FUELS

For an understanding of how heat is to be provided for our houses it is convenient to begin by examining how heat—from whatever source—is distributed within the house; for some fuels are better suited to certain methods of heat distribution than are others.

In a central heating system a single heat producer serving an entire dwelling (commonly, a furnace of some kind) directly heats air or water, or produces steam, and distributes it throughout the living space. Such systems are likely to be more expensive to install in houses already built than they are to install in houses under construction: but they make up for their costs in convenience and effectiveness.

All automatic central heat furnaces rely on electricity for some part of their function. Electric pumps or belts bring fuel to the furnace's combustion chamber; electric thermostats regulate the rate of combustion or turn the furnace off and on; electric fans or pumps distribute the furnace's heat.

In a **hot water heating system** which is separate from the hot water plumbing supply, water, heated in a boiler, is circulated through pipes and radiation units to warm rooms by simple radiant heat. The water then returns to the boiler to be re-heated and sent on its circuit again.

Modern versions of this system use an electric pump to enhance circulation of heated water. The water, however, will circulate to a limited extent without being pumped as long as it is hot. These systems, therefore, will continue to function in a power failure— provided, of course, that the furnace does not depend on electricity to operate in the first place. To make self-contained hot water heat systems more immune to freeze-ups in the event of a fuel failure or extended power outage, some householders, since the early 1970's, have been adding an antifreeze preparation to the re-circulating hot water supply. However, the use of antifreeze in these systems has not been endorsed by all professional heating contractors.

A **steam heat system** uses the same equipment that hot water heat requires, but water in the boiler, heated by the furnace, is distributed to room radiators as steam. Steam heat is seldom installed today, for it is more difficult to regulate room temperature evenly in a steam-heated house than in houses heated by other means.

In a **warm air central heat system**, fire in the combustion chamber heats air in a second chamber, the heat exchanger. The warm air is then conveyed from the heat exchanger throughout the house by means of ductwork and registers. The heat exchanger arrangement keeps smoke from the fire from entering the ducts and getting into the house. Cool air is drawn into the heat exchanger from cold air intakes placed at a distance from the warm air registers. Warm air usually is forced through the ducts by an electric fan in the heat exchanger. However, it is possible to heat several rooms more or less directly above the furnace by means of "gravity" warm air: the natural tendency of the warm air to rise carries heat through a main duct to a central area. Such convected heat has no fan and few ducts, and in the northern snowbelt it can heat only a few rooms adequately.

Most warm air heating systems dictate that domestic hot water be heated otherwise than by the furnace. The reason is that there is virtually no control over the amount of heat that water coils or water fronts are subjected to in the combustion chamber of a warm air furnace. On a bitterly cold day, when your furnace is running constantly, steam would come out of your faucets if the furnace also heated your house hot water.

The use of oil—Number 1 (kerosene) or Number 2—for home heating has declined somewhat in the last few years, although oil is still in wide use in rural New England. Its per-gallon cost has increased drastically since about 1970, and promises to continue its rise.

Number 2 Fuel Oil is a little less expensive than kerosene, and has slightly more heat value (that is, a given volume of Number 2, burned, produces more heat than the same volume of kerosene —see Heat Value Table). Number 2 is also thicker than kerosene, however; unless it is stored inside it may, in cold weather, become too thick to flow easily. Kerosene therefore is usually used to heat dwellings (mobile homes, for instance) where there is no room inside for fuel storage; and it is also burned in small space heaters (see Chapter 4).

Professional annual servicing of oil burners, in which oil lines, filters and valves are cleaned, and electrical contacts are checked for corrosion, are important in keeping systems running safely and economically.

Sixty percent of all American homes are heated by **natural gas.** Scarce in rural areas, natural gas (methane), piped through service mains, is the house heating fuel used by most city and town dwellers. In the country centralized gas distribution systems are not practicable, and the gas used for home heating is **Liquid Petroleum Gas,** usually propane (called LP gas). LP gas is held under pressure in liquid form in tanks kept outside the house, and turns to gas when it is released from the pressurized tanks to enter the house. It is heavier than air, and has more than twice the heat value of natural gas (see Heat Value Table). LP gas is generally more expensive than natural gas, and it is used for cooking and hot water heat in country houses more than it is used for heat.

Gas heating systems—natural or LP—require little maintenance. Pilot lights and safety thermostatic controls which shut off the gas when the pilot light is out are at the heart of these systems. All gas systems should be serviced and adjusted by professionals in every case: there is little that the householder can safely do in the way of attending to them.

Although the use of coal for home heat has declined over the past thirty years, increasing costs of gas, oil and electricity, and the presence of vast coal reserves in the United States, may be factors in a re-birth of coal-burning.

Where it is still available, coal is usually much cheaper than the other standard sources of central heat: oil, gas and electricity. Coal is dirty, however, and requires large storage space, as the others do not. A more important factor in the demise of coal as a major home heating fuel has been that coal fires do require tending, even those in furnaces with automatic stokers; therefore, coal is less convenient than more costly heat sources.

Coal for home heating may be either hard (anthracite) or soft (bituminous). Hard and soft coals have similar heat values (see Heat Value Table), but soft coal burns dirtier than hard, producing more ash and soot. Hard coal comes in a variety of sizes, depending on the kind of equipment that will be used to burn it. In furnaces equipped with automatic stokers, "rice" coal is often used. Rice-grain-sized bits of coal are fed to the fire by a conveyer belt from the stoker—a box which must be kept full of coal. The stoker is activated automatically by an electric thermostat or timer, to fuel the fire when a pre-set minimum temperature is reached, or after a pre-set time has elapsed.

For hand-stoked coal furnaces larger pieces of coal are available, including "pea" and "chestnut" coal. Cannel coal, a soft variety, used to be burned in thousands of homes in open fireplace grates; cannel coal usually comes in fist-sized pieces. All coal sizes produce the same amount of heat.

Starting a fire in a coal furnace or space heater usually involves building a small fire of charcoal or wood kindling and then adding coal gradually until the fire is well established in the coal. Never use chemical fire-starters—the sort sold for outdoor barbecue—in a furnace or space heater. Coal furnaces and space heaters are equipped with draft controls: one or more adjustable

HEAT VALUE TABLE

Fuel	Btu Output of a Convenient Unit	Amount Needed to Produce Btu Equivalent of 1 Ton of Hard Coal	Amount Needed to Produce 100,000 Btu (1 Therm)
Soft Coal	26 million Btu/ton	1 ton	.03 ton
Hard Coal	25 million Btu/ton	—	.04 ton
Wood	24 million Btu/cord	1.1 cord	.05 cord
Number 2 Oil	14 million Btu/100 gallons	183 gallons	.70 gallon
LP Gas	4,000 Btu/cubic foot	6,200 cubic ft.	24 cubic feet
Electricity	3,400 Btu/kilowatt hour (kwh)	7,500 kwh	30 kwh
Natural Gas	1,000 Btu/cubic foot	2,500 cubic feet	97 cubic feet

All numbers are rounded off.
1 Btu (British Thermal Unit): defined as the amount of heat required to raise the temperature of one pound of water by one degree Fahrenheit.

openings in the firebox which allow the amount of oxygen reaching the fire—and thus the speed and heat-intensity with which the fire burns—to be regulated. After adding coal to a fire in the course of starting up, or in adding coal to an existing fire, it is necessary to let the fire burn for a few minutes with the draft control open. This lets the fire get started and, more important, it gives potentially hazardous gases released as coal begins to burn a chance to burn off harmlessly. Failure to let a coal-burner "gas off" properly after fueling can result in an accumulation of carbon monoxide, which is lethal if inhaled. Other gases released by burning coal can explode if they are allowed to build up in a coal furnace or heater.

In coal furnaces and space heaters, the burning coal rests on a grate which must be regularly freed of ashes. Most coal-burners have a handle outside that is used to shake the grate while the coal is burning. The ashes fall off the grate into a drawer below from which they are emptied. It is important to keep the clean-out drawer empty: don't let ashes build up in it until they touch the grate.

Hand-stoked coal furnaces can be run on less coal than automatic-stoker furnaces. There is some art to holding a coal fire in a furnace (or space heater) for long periods of time, but practiced fire-tenders can heat a house all day on two fuelings, one in the morning and one at night. This makes coal a convenient fuel for people who must be away from home all day at work and don't want an oil- or gas-burner—or an electric heat system—running wastefully when the house is empty.

Electricity is easily the most convenient, cleanest, safest way to heat your house—unfortunately it is also the most expensive way in many areas. Electric central heat requires no fuel storage space, no fuel handling. Because nothing is burned in your home, there is no need to arrange for venting an electric system.

Most electric central heat systems use electrically heated cables or other heat elements built into ceilings, baseboards, or wall radiators. A different type of electric system uses a "heat pump"— essentially an air conditioner run in reverse to collect warm air from outdoors and blow it into the house in exchange for cool air which it pumps out. Heat pumps are more common in milder climates than in the northern snowbelt.

Two other potential sources of central heat are wood and solar energy. The climate of the northern snowbelt, and the high costs of extensive solar heat installations, make the second impracticable for any but auxiliary heat use over most of the north (see Solar Heat, below). Central heating with wood is possible where fuel wood is abundant, but it involves large expenditures of time and effort on the part of the householder.

Wood-burning furnaces for central heating were once fairly common, at least in northern New England, and they are coming back into use now after forty years of neglect, as other methods of central heating become more and more costly. The old wood furnaces were just big cast-iron fireboxes with an air intake and a cold air return. New designs allow for more complete combustion of wood through adding another air intake ("secondary air"). They also can be hooked up to electric thermostats, with which some wood furnaces function as well as oil, gas or electric systems. Most wood furnaces can be easily adapted to provide hot water as well as house heat, thus reducing home energy costs in two ways.

A major problem in wood-burning, with whatever kind of equipment, is the build-up of combustion by-products, loosely called "creosote," in stovepipes and chimneys. Creosote deposits in stacks can catch fire. They may burn away harmlessly, or they may burn your house down. The more wood burned in any furnace or stove, the more likely it is for dangerous amounts of creosote to accumulate quickly. Central heat wood furnaces, because they burn large quantities of wood, often at a relatively slow rate, produce creosote abundantly. New furnaces, by providing secondary air and therefore more complete burning, develop creosote problems to a lesser degree than do older furnaces. Nevertheless, creosote build-up, necessitating shutdowns for chimney cleaning, remains a drawback to wood central heating.

Another drawback is the amount of wood required to heat an entire house from a single furnace. One model of wood furnace on the market today can take a quarter of a cord at a single fueling. A moderately large house in a cold region might well burn a dozen cords a winter in its wood furnace, and most householders are not prepared to handle or store wood in those quantities. Creosote problems and wood handling problems, plus high prices ($1,000 and up) and installation costs, rule wood central heat out as an option for most householders.

For those who want the savings on fuel that may come with burning wood in a central heating system, but don't want to rely on wood exclusively, a choice of **combination-fuel furnaces** is available. Wood-oil, wood-gas, wood-coal and wood-electric systems are sold. Different fuels are burned in separate combustion chambers (see drawing). In most combination-fuel furnaces, the more expensive fuel is burned as a back-up for wood: the oil-burner, for example, starts up automatically on a thermostatic control when the wood fire dies down. Like one-fuel furnaces, combination models can be made to work with any kind of heat distribution system (forced warm air, hot water, etc.). Combination-fuel furnaces have high first costs ($1,500 and up), and may be expensive to install, for instance, if a chimney must be built for the wood-burning element.

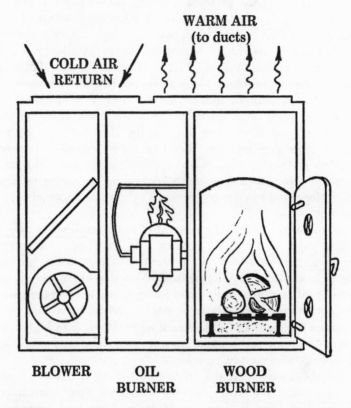

WARM AIR
(to ducts)

COLD AIR
RETURN

BLOWER   OIL   WOOD
              BURNER   BURNER

Diagram of wood-oil combination fuel furnace for central heat.

WOOD HEAT

Wood as fuel has a heat value comparable to that of other fuels
(see Heat Value Table), but it is not a standardized, uniform
product, as other fuels are. There are also differences in heat
value between wood from different species of trees, and these are
differences the wood-heat user should be aware of.

Relatively dense and heavy wood makes better fuel than less

dense, lighter wood. In determining the heat value of a particular load of wood, however, its dryness must be taken into account as well as its density. When wet wood burns, the water in it goes up the chimney as steam. The heat of that steam is lost. As much as two-thirds of the potential heat energy in a volume of wood can be lost if the wood is burned wet. From the point of view of wood dryness, therefore, the best firewood is that with the lowest moisture content when green, for it will need less drying time before it can be burned and will dry more thoroughly in most circumstances. The ideal fuel is a dense wood with relatively low green moisture content.

The denser-wooded tree species in the northern snowbelt are broadleaf deciduous trees, while lighter wood generally comes from evergreen species which bear needles. For density and—usually—for low moisture content, wood from broadleaf deciduous trees makes much better firewood than wood from evergreen needle-bearers, and should be preferred whenever either can be had. Exceptions to this rule are the deciduous tree species having wood that is light and makes poor fuel: poplar, sumac, soft maple and others.

FIREWOOD QUALITY OF TWENTY-FIVE NORTHERN TREES

Very Dense (Best Firewood)	**Medium Dense** (Good to Fair Firewood)	**Light** (Fair to Poor Firewood)
Hickory species	Ash species other than White	Hemlock
Locust species	Paper Birch	Pine species
Oak species	Larch (Tamarack)	Spruce species
American Beech	Elm species	Aspen (Poplar)
Apple	Maple species other than Sugar	Butternut
White Ash	Black Cherry	Cedar species
Sugar Maple	Sycamore	Cottonwood
Yellow and Black Birch	Sassafras	Basswood
		Fir species

Species arranged in *approximate* order of density (densest at column tops).

Disadvantages of wood heat are: wood requires a great deal of storage space, preferably under-cover storage space; wood fires require relatively frequent attention; chimneys in houses that burn wood require regular cleaning; wood heating presents a greater fire hazard than other methods. Furthermore, the time and labor involved in getting your own wood in any quantity are really considerable, for the wood must be found, cut, split, stacked and seasoned, in a series of steps which should stretch over a year if green wood is to be properly dry when it is burned.

If you expect to cut your own wood from the forest, learn the tree species in your area so you will be able to select those that make the best firewood. Allow plenty of time for your wood to

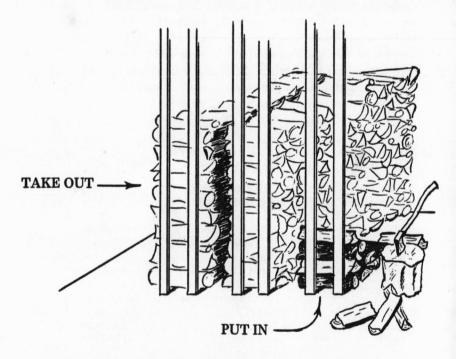

TAKE OUT ⟶

PUT IN ⟶

Wood-pile layout to promote drying and avoid having the last wood added to pile become the first wood taken out.

dry before use: it is good practice to cut in the fall and winter the wood you will burn the *following* winter.

Wood-cutting tools are simple and—with the exception of chain saws—inexpensive. You will need a saw with a replaceable blade (bow or buck saw), an axe, steel splitting wedges and a sledge-hammer, and possibly a splitting maul (this last looks like a light sledgehammer with one striking face tapered to an axe blade). Learn the care and safe use of these tools, and, if you are a beginner with any of them, take it slow at first. Avoid using the axe, hammer or maul when you are tired.

Chain saws reduce significantly the time required to prepare firewood. To compensate for their convenience, chain saws cost a lot ($100 and up, usually), need expensive service, consume gas and oil. They are also loud, smelly and potentially lethal. If you cut less than two or three cords of wood a year a chain saw may not be worth your while. Shop around carefully if you do decide to buy one, for chain saws are available in a bewildering array of sizes and types. The best way to choose a saw is often to settle on whatever make is sold by a local dealer whom you know to have a good reputation for servicing the saws he sells.

Where to cut presents a major problem for anyone who doesn't own a woodlot. Ask property owners in your area for permission to cut dying, malformed or overcrowded trees for firewood; many won't agree, some won't mind, and some will be glad to have you do the work. In areas where timber-cutters have been harvesting trees for lumber, unusable tree tops and branches lying on the forest floor can be cut for firewood. Indeed, owners are often happy to have this done, for the dead tops are an obstacle to getting about in the woods, and constitute a fire hazard. Finally, areas in some state forests are sometimes opened periodically for selective firewood cutting (ask the state forest authority in your area).

If you buy firewood, remember that it is measured and sold by the *cord,* and therefore most practical calculations of the costs involved in heating with wood as compared to other fuels depend on the cord unit. A cord is a stack of wood four feet high, four feet deep and eight feet long. Although firewood dealers price wood by the cord they often don't handle or deliver wood in even cords. In practice, you get a truckload of wood, and that truckload may contain more or less than the cord of wood you thought you were buying, depending on the size of the truck and on the size and shape of the pieces of wood in the load. Few firewood dealers will wait around for their money while you carefully stack the wood they have just delivered to see if it makes a full cord. If you have not bought firewood in your area, you do well to go to a dealer who is known to your neighbors (and perhaps to your county forester or agricultural extension service) as one who delivers the quantity ordered.

Another pitfall in buying firewood is opened by the dryness of the wood you buy. As we have seen, firewood should be thoroughly dried, or seasoned, before it is burned. What is seasoned wood to the seller may not be seasoned wood to the buyer. "Seasoned" *should* mean air-dried under cover for at least eight months, but it often means to the dealer simply "not green." Cut or dead wood which has been lying out in the weather burns better than green wood (unless it is rotten), but it is a far cry from properly seasoned wood. Unfortunately it is not always easy to tell well dried wood from wood that is not green but not dry, either. Again, the buyer's best course is to go to a dealer whose wood is known for its high quality.

Learning the trade secrets of firewood dealers who are no better than the rest of us is not the only way to intelligent wood-buying. When to buy is even more important. Buy wood in late spring and summer when demand for it is low and its price is down. Buy green wood if it is cheaper than dry, and if you can let it dry for a sufficient time before you will use it.

AUXILIARY HEAT AND HEATERS

The vast majority of existing houses in the snowbelt area are equipped with central heat of one kind or another, and few householders would replace these systems with more economical alternatives, if they could. Short of doing away with central heat entirely, however, it is almost always feasible to use space heaters as auxiliary heat sources, supplementing central heating systems or replacing them at convenient times. Limited use of space heaters which run on relatively inexpensive fuels like wood can often effect savings by cutting down on one's use of the more costly fuels required by most central heating systems.

Except in emergency situations (see Chapter 4) fuel oil (including kerosene), gas and electricity are not practicable as sources of auxiliary heat to supplement central heating systems. On the contrary, oil, gas and electricity are precisely those heat sources which use of auxiliary heat sources is intended to replace.

Just because coal usually costs less than other central heat sources, however, it is well suited for use in an auxiliary heat plan. Coal space heaters are on the market, most of them European imports. These small coal-burners usually burn hard coal, in "pea" or larger sizes. They are similar in size and appearance to wood-burning stoves, and have installation requirements similar to those of wood stoves (see section on Wood Heat). Coal and wood are not interchangeable fuels, however; wood shouldn't be burned in coal heaters, nor should coal be burned in stoves intended for wood.

In areas where firewood is expensive and/or hard to get coal space heaters may make excellent auxiliary heat sources supplementing oil, gas or electric central heating. Especially in suburban areas, coal may be a logical choice for auxiliary heat. In these areas, where the most common housing unit is the single family house, cellar space for coal storage will often be ample; and coal will be available from the nearby city. Firewood, on the other hand, will be in short supply: it won't be there for the cutting in the immediate neighborhood, and the cost of trucking it in from the country will make it expensive.

As mentioned in the discussion of coal as fuel for central heat, coal fires produce **carbon monoxide gas,** which is lethal if inhaled, even in very small amounts. Coal space heaters entail a greater risk from carbon monoxide than coal furnaces, since heaters are usually located in the lived-in rooms of a house. Always let a coal fire run hot, with the heater's draft controls open, after you start it; and shut down draft controls only after carbon monoxide has had a chance to escape. Never sleep in a room with a coal fire if you can avoid it. Finally, remember that carbon monoxide, though particularly associated with coal fires, is produced by incomplete combustion of any carbon-containing substance, including kerosene, natural gas, gasoline and wood. Heaters which burn any of these fuels should only be used in well-ventilated spaces. Fires burning in ill-ventilated spaces, in addition to giving off carbon monoxide, can draw oxygen from the space, producing dangerous shortages of breathable oxygen in extreme circumstances.

Where it is abundant, wood is a fuel well suited for use as auxiliary heat. A recent survey showed that more than half the houses in Vermont have a wood stove in use, and similar statistics could probably be developed for much of the Northeast.

To use wood efficiently for space heating you need a closed stove. Fireplaces, masonry or metal ("Franklin fireplaces") do not produce much usable heat. In fact, one study I have read of showed that an open fire in a fireplace *cooled* the house; the fire in effect pumped warm air out of the house and up the chimney. There are special grates, screens and duct systems on the market which can adapt a fireplace to put out more room heat. Most probably help, but none makes a fireplace a substitute for a good closed stove.

At this writing there are at least 150 different types and models of wood-burning space heater on the American market, produced by at least fifty different manufacturers in the United States, Canada and Europe. Each manufacturer promotes his own product with claims of superior heat output through novel design or use of materials. It is difficult to know how to assess these claims, but there are points that the prospective wood stove user should keep in mind.

The heat output of a stove is roughly proportional to the size of that stove. Material, design, color, the wood you burn, all make some difference, but size is the key: a big stove makes more heat than a little stove.

Materials are not that important to a stove's heat output,

though they affect other important aspects of the stove's use, notably its longevity. Most stoves are made of cast iron, sheet steel or steel plate. All these have similar heat conducting properties, and so all have similar heat outputs. Cast iron and steel plate last longer (and are more expensive) than sheet steel. Between cast iron and steel plate, however, there seems to be little to choose.

One important difference in design among stoves is that some are airtight and some not. In an **airtight stove** the only ready access of air to the interior of the firebox is via an adjustable air intake port or ports. It is possible to control the rate of burning in an airtight stove more effectively than in a non-airtight. This makes airtight stoves more efficient than others: they use less wood to produce a given amount of heat, and therefore need less frequent tending.

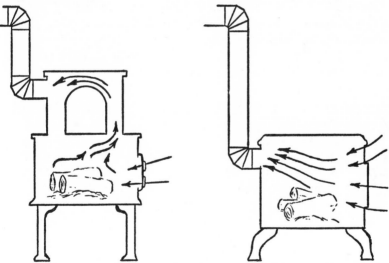

Air flow in airtight (left) and non-airtight wood stoves. In the airtight model, air access is only through draft control openings on door (arrows); in the non-airtight, air can enter around the door and at structural joints. (Upper arch in airtight stove is intended to increase heat output in some stove models by holding smoke in the stove—arch is not an integral part of airtight design.)

Another design feature of some wood stoves is a thermostatic control. Stoves with thermostats are airtight. In these stoves the closure of the main air intake is controlled automatically by a heat-sensitive mechanism which reduces air flow to the fire when a pre-set temperature has been exceeded, and restores air flow below that temperature. Most thermostatic stoves do offer some degree of temperature control, but they have the drawback that they tend to shut down, slowing the rate of the fire, unless their air intake ports are kept wide open at all times (which makes the thermostat feature irrelevant). As a result of their fostering slow, relatively cool fires, thermostatic stoves often build up creosote in pipes and chimneys more than stoves without thermostats.

Finally, as you read the claims of rival stoveworks for greater and greater Btu outputs or cubic foot heating capacities, remember that the heat output of your stove is not the only factor which determines how warm it will keep you when winter comes. If your house does not have adequate insulation, especially, then your corners will be cold no matter how hot your stove is.

The proper installation of a wood stove is vital for the safety of your home. If possible, have your stove hooked up by someone who has experience in the work. Whoever does the installation, make certain the stove and pipe are placed at a safe distance from walls, floors, furniture, or anything else that can burn. An eighteen-inch clearance from floors, and a two-foot clearance from walls and furniture are almost always required. Stoves should not be placed on bare floors: put an insulating asbestos pad or a shallow box of sand under your stove. Avoid long lengths of pipe between your stove and the chimney, especially long horizontal lengths: the longer the pipe the more the smoke from the fire will cool before it reaches the outside air, and the more combustible creosote will be deposited in the pipe and chimney. A six-foot pipe and two bends or elbows are the ideal upper limit. Avoid running stovepipe through interior walls. If the pipe must pass through a wall make sure the passage is insulated: an insulated opening three times the diameter of the pipe it will have to accommodate is recommended. Materials for installing wood stoves

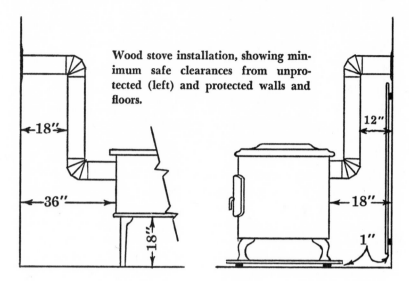

Wood stove installation, showing minimum safe clearances from unprotected (left) and protected walls and floors.

safely are readily available at stove dealers and hardware stores in areas where wood heat is in wide use.

The condition of your chimney has a great deal to do with how efficiently and safely your wood-burning equipment operates. Chimneys clogged with creosote not only present a fire hazard, they also prevent your furnace or stove from burning properly. Creosote is deposited in chimneys as smoke cools with distance from the heat of the fire. In particular, the water component of smoke, in the form of steam, holds combustion by-products in suspension. As the smoke cools the steam condenses and the by-products are left behind—often in the chimney. It follows that burning green or wet wood produces more creosote than burning dry wood, and is to be avoided if there is any recourse at all. Even if your wood is dry, though, your chimney and stovepipes will accumulate creosote over time, and will need regular cleaning.

In order to effectively draw off smoke from a fire and allow fresh air to reach it a chimney must be warm. Therefore the best chimneys are ones which warm up quickly and stay warm, especially in segments exposed to the cold outside air. Ideally, chimneys are made of brick, concrete block, cement or fieldstone-and-mortar. All these materials, and the men who work with them, are expensive. Less expensive are steel chimneys which can be put up quickly by almost anyone. These prefabricated metal chimneys are insulated (in effect they are a little stovepipe inside a big stovepipe), allowing them to hold the maximum amount of heat.

Most authorities advise against connecting more than one stove, furnace, or other appliance to the same chimney flue. Doing so can cause the air flows from the separate appliances to interfere with each other, resulting in smoking and perhaps in increased creosote build-up. If you cannot avoid connecting two or more appliances to the same chimney flue, make sure the flue is wide enough to accommodate them (consult a builder or fire inspector), and put their entrances into the flue at different heights. Remember that if you install a wood stove to supplement your central heating system, and your chimney has only one flue, then you have a situation in which more than one appliance is connected to that flue: for your central heat furnace will probably vent into the flue as well as your new wood stove.

Chimneys should be cleaned at least twice a year, once just before the onset of winter and once in mid-winter (stovepipes should be dismantled and cleaned every month in winter). Modern chimneys have sectional tile flues inside the masonry exterior, but older chimneys often do not. Chimneys with unprotected flues require more thorough cleaning than others, for a fire in such a chimney can easily spread to the house through spaces between bricks in an exposed flue.

Chimneys can be cleaned by lowering large stiff brushes, chains or weighted sacks into them on ropes from the top and pulling up and down. A rope on each end of whatever is doing the cleaning, and two people—one on the roof and one below—make the job easier. Don't use your vacuum cleaner: the soot can damage it.

A less strenuous way of keeping your chimney clean is to use chemical chimney cleaners. These are sold in powder form. A measured amount of powder is thrown on a hot fire at regular intervals, as directed, during the heating season. Fumes from the burning powder react with creosote deposits in the chimney, causing them to flake off. Many people doubt the effectiveness of chemical cleaners, and feel they can damage bricks and mortar. Whatever their pros and cons, these products certainly should not be used in place of careful manual chimney cleaning.

Professional chimney sweeps can be found in most areas where wood or coal are widely used for heat. They have special equipment, including extra-heavy-duty vacuum cleaners, and are worthy of their hire for most householders.

Fire-tending deserves mention, for any kind of wood heating equipment requires more attention from its owner than equipment using other fuels. You will have to get the hang of starting and keeping a fire in your stove or furnace. In lighting a fire, start small with kindling wood only, and add full-sized pieces of wood only after the kindling fire is well established and starting to wane. This allows the stove, pipe and chimney to warm up thoroughly, promoting a better draft and hence a better fire. Never burn trash in a stove. Chemicals released from burning some materials—notably plastics—attack metal and can damage your stove and pipe.

SOLAR HEAT

Theoretically, all heat from burning fuel is solar heat. When we burn oil, coal, gas or wood—or use electricity produced by generators which burn them—we are retrieving heat energy latent in vegetable matter, and that energy came originally from the sun. Nevertheless, despite similarities between solar heat and other forms, there is one fact about solar heat which makes it unique. It's free. Sunlight costs nothing, either in money, labor or time. The only investment required by a solar heating system is in equipment to collect, store and distribute the sun's heat.

Solar equipment is expensive, however, and in many areas of the northern snowbelt there may not be enough sunlight available in the winter to make extensive solar heating practical, given today's solar technology and costs.

"Active" solar heating systems, so-called—those using special solar collecting and storage equipment—have three components: solar collector; heat storage facility; heat distributor. In a solar collector, a light-absorbing surface is exposed to the sun, which warms some medium inside the collector. In many solar collectors this medium is water, protected from freezing inside the system by being mixed with antifreeze solution. The water, heated in the collector, circulates to a storage tank which may also serve as a radiator for distributing heat. In other types of solar heat systems, air is warmed in the collector and its heat stored in an insulated container filled with small rocks. Heat may then be circulated from the rocks by a fan through ducts, as in a conventional warm air system.

Simple solar collectors are flat, glass-topped boxes containing a system of pipes or tubes through which water or air is circulated. The inside of the box and the pipes are painted black, and the box itself is insulated to retain heat. These so-called "flat-plate collectors" are often built on house tops or walls, and tilted and oriented to receive maximum sunlight.

Active solar heating systems, to be practical in northern latitudes, require large solar collector surfaces and heat storage volumes. The book, *Designing and Building a Solar House*, by Donald Watson, estimates that a "large-capacity" solar heat system, installed, could cost as much as $8,500 in 1977. This figure represents the cost of installing such a system in a new house. Adding an extensive solar heat system to an existing house not designed with solar equipment in mind could be far more expensive, and is seldom feasible. In the northern snowbelt, moreover, even a house equipped with a large-capacity solar heat system would almost certainly need a conventional heating plant as well, for back-up use in long periods of low sunlight, adding to the cost of house heating equipment of all kinds.

Although total reliance on the sun for home heating, in the north, involves costs which are out of reach of most householders, the use of solar heat on a limited scale, to supplement conventional heating systems, is often quite practicable. Through the use of double- or triple-glazed window panes, skylights, sun porches, and south-facing windows can easily be made to admit appreciable amounts of heat. In improving the heating abilities of windows, however, it is crucial to provide tight blinds to close over them, inside; much radiant heat can escape through window glass at night.

A more ambitious, but still simple, way to use the sun to provide auxiliary heat is to install a solar water heater to pre-heat your hot water. Pre-heating can cut hot water costs significantly, and simple, low-capacity solar water heaters are more practical for northern houses than elaborate systems. In one simple solar hot water set-up, water heated in a flat-plate collector rises (without benefit of pumps in a properly designed system) to a storage tank, whence it is piped into your standard water heater. In a solar hot water heater, antifreeze, which is toxic, cannot be added to the water directly to keep it from freezing. Therefore in these systems the water–antifreeze mixture heated in the collector fills a tank in which is a second sealed tank or coil. Potable hot water is held in this second tank and kept from freezing by the water–antifreeze mixture in the larger tank (see drawing).

The cost of a simple solar hot water system like the one described could be as high as $1,500 if all equipment was bought and the installation was done by professionals, or as low as $500 if equipment was homemade when possible and installation was done by the householder.

For anyone contemplating installing active solar heat equipment of any kind—including simple hot water systems—it is important to remember that the solar equipment industry is in its infancy in the United States today. Organization of the manufacture, distribution, marketing and installation of this equipment is still uncertain in most of the country, including the northern snowbelt. Therefore it is crucial to take care before investing in a solar heat system, however modest. Look around carefully, and get good advice. In one study of the performance of solar hot water heaters in Massachusetts, the generally disappointing results showed the most important failings of the systems

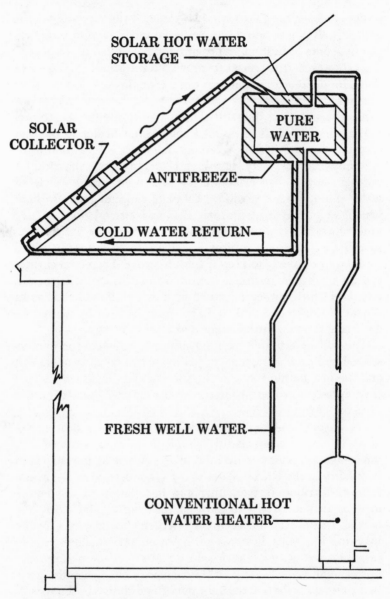

SOLAR HOT WATER
STORAGE

PURE
WATER

SOLAR
COLLECTOR

ANTIFREEZE

COLD WATER RETURN

FRESH WELL WATER

CONVENTIONAL HOT
WATER HEATER

Diagram of simple solar water pre-heater.

were poor quality-control of manufactured solar equipment, frequent break-downs, and bad installation jobs. It wasn't the sun who fell short.

The Federal Government's Energy Program has promised tax rebates, or other incentives, for householders who install solar heat equipment. At this writing, however, no incentives have been settled upon or approved by Congress—a fact to remember as you deal with self-proclaimed solar heat contractors who want to persuade you to avail yourself of their services.

COUNTING COSTS

In determining which source of home heat gives the most comfort for the smallest expenditures of money, time and labor, a number of factors must be taken into account. Local availability and costs of various fuels, and the amount of fuel needed to heat your house over a winter, are the most obvious considerations. Also to be weighed are: first costs of heating equipment (purchase price of furnace, stove, etc.) ; labor costs (for installation of systems) ; interest rates (if you borrow money to purchase heating equipment) ; projected future increases in fuel costs in your area; depreciation on equipment; hidden costs of "free" firewood (for example, the cost of a chain saw and gasoline to operate it) . In devising an auxiliary heat plan, especially, which will balance costs of two or more fuels to effect savings, determining possible savings can be a complicated matter. Apparent savings can be deceptive.

Wood space heaters used to supplement central heat can always produce significant short-term gross dollar savings *if you cut your own wood*. If you buy firewood, however, even if it seems to cost far less than the fuel used by your central heat, you will find your savings cut drastically. To take a simplified example, suppose your central heating fuel is oil (cost = .40 per gallon) and you burn 1200 gallons in a winter, over six months. Your total oil bill is $480 over twenty-four weeks, or $20 per week. Now suppose you formulate an auxiliary heat strategy based on wood in which you substitute wood for oil heat each evening (a natural time to do so in many homes, when the heated area of the house can be reduced with minimal discomfort and inconvenience) . Let us say you burn wood in preference to oil for four hours a day. You are using wood instead of oil for twenty-eight hours a week or four weeks (672 hours) over the entire winter, and the total cost of the oil you have conserved is $80—your *gross* dollar savings in fuel. Suppose you have burned two cords of wood in that same winter, and suppose wood in your area is relatively cheap—$30 per cord. Your net savings in fuel costs equals only $20, and you haven't yet figured in the cost of your wood stove, which, even if you amortize it over several years will subtract further from your total savings.

GOOD READING ON HEATING

Gay, Larry. *The Complete Book of Heating with Wood.* 1974. Garden Way Publishing Company, Charlotte, Vermont. Available in bookstores or from the publisher for $3.95. *Good introduction to wood as fuel; survey of wood stove types.*

Shelton, Jay and Shapiro, Andrew B. *The Woodburners Encyclopedia.* 1977. Vermont Crossroads Press, Waitsfield, Vermont. Available in bookstores or from the publisher for $6.95 (paperback) . *Thorough and thoroughly excellent on all aspects of wood-burning; includes inventory of available stoves and furnaces, with specifications.*

Simons, Joseph W. *Home Heating: Systems, Fuels, Controls.* 1975. Published by the United States Department of Agriculture (Farmers' Bulletin No. 2235) . Available free from the U.S. Government Printing Office. *Brief survey of conventional central heat fuels and equipment.*

Watson, Donald. *Designing and Building a Solar House.* 1977. Garden Way Publishing Company, Charlotte, Vermont. Available in bookstores or from the publisher for $8.95. *Exhaustive, very detailed treatment of solar heat for homes, with estimates of its economic feasibility today in different regions.*

4 *Emergencies of Winter*

HOWEVER EFFECTIVELY you have prepared for winter there will be times when the weather outflanks your defenses and forces you to deal with winter more on its own terms than you would wish. Power failures, frozen pipes, and snowed-in houses are a few of the rural winter episodes which will drive you to emergency measures—though I hesitate to speak of these situations as emergencies, for they are commonplaces of winter in snow country. Special preparations are also necessary if you leave your house unoccupied in winter for any length of time.

POWER FAILURES

Power failures are a usual part of winter in the country, where electrical service lines are strung up on poles, exposed to ice, high winds and falling tree branches, rather than buried under streets as in the city. Your house may be without electricity for several hours—or for several days. In a long power failure the amount of inconvenience you suffer depends on how well you are equipped to get along without electricity. A prolonged power failure can be a calamity, or it can be, almost, an adventure.

A power failure knocks out all house systems that require electricity: electric lights, stoves and food freezers go out; electric pumps that furnish running water stop; electric thermostats, fuel pumps, furnace fans and hot water heaters don't work. Your ability to weather a power failure depends on your ability to provide light, heat, and water for your house without using electricity from outside. If you have an emergency generator this presents no great problem.

Home generators, powered by gasoline engines, can be had in a range of sizes. A five-horsepower generator, rated to produce 2,000 watts, will run your house lights and your furnace for about two hours on a half-gallon of gas, and costs about $300. Bigger generators can supply nearly all your house current (until you run out of gas) and are correspondingly more expensive.

If you have no generator or other single substitute source of electricity, you will need, first, light, from candles, flashlights, lanterns or kerosene lamps of one kind or another.

For anyone who can expect power failures that are any more than both few and short, candles are not adequate as emergency light. Too many are needed to make useful illumination, and, though candles are probably no greater a fire hazard than kerosene lamps and lanterns, the fact that you must use relatively many of them multiplies their danger.

Kerosene lamps and lanterns are better illuminators than candles. They require care, however, and they require you to have on hand a supply of kerosene (which is highly flammable and should be kept in a special safety can, outside the main house if possible). For maximum light production lamp wicks should be kept trimmed and glass chimneys should be cleaned. Most kerosene lamps burn for ten to twelve hours on a quart of fuel. Aladdin brand lamps, in which an incandescent element (the "mantle") increases light output from the burning wick, produce more light than conventional lamps, but the flame must be adjusted carefully in these lamps to avoid charring the mantle (kerosene lamps other than Aladdin models may use highly polished reflectors to increase usable light output). One of the several Aladdin lamp models available can produce the light equivalent of a seventy-five-watt bulb, and can burn for twelve hours on a quart of kerosene. Larger models produce more light. All Aladdin lamps are relatively expensive ($20–$75).

When not in use, kerosene lamps may be dismantled and the fuel reservoirs stored tied up in plastic bags, or fuel may be emptied from lamp reservoirs into sealable bottles for storage.

This saves kerosene, which can evaporate quickly if left exposed to the air. For the safe use of any kerosene lamp or lantern it is important never to use any fuel but kerosene: no gasoline, no lighter fluid.

Emergency **lamps that do not use kerosene** include propane-burning models, in which a replaceable propane gas cylinder is attached to the burner; and Coleman lanterns, which burn the vapor of a special, highly refined liquid fuel, white gas, which is held under pressure in the fuel reservoir. Both these types of lamp produce very satisfactory light, at least as bright as that of Aladdin kerosene models.

Flashlights and battery-powered lamps may be a better solution to the emergency lighting problem than kerosene or other fuel-burning lamps. Battery-powered lamps are expensive, but not more expensive than Aladdin kerosene lamps, and the advantage of doing away with fuels and open flames is a considerable one—especially in houses with children. A disadvantage of relying on battery-powered emergency lighting is that—unlike lamps and candles—batteries cannot be stored indefinitely until they are needed: they lose power on the shelf, slowly but surely.

Flashlights and kerosene lamps are the basic equipment needed to meet a prolonged power failure, for you will often need light to carry through measures for providing emergency heat and water. A battery-operated radio is also a help, for it will enable you to keep track of weather forecasts and local reports, and so to guess how long your power will be out.

If the first need in a power failure is for light, the most pressing need is for **heat**. If you can heat your house, or any part of it, during a power failure, you can last the episode out with more or less inconvenience; if you cannot, you may have to evacuate your house for the duration of the power failure, and you run the risk of water damage from frozen pipes that burst. An auxiliary heat source of some kind—a fireplace, a coal- or wood-burning stove, a gas or oil space heater, whatever—which may be kept as an economy in ordinary times, becomes a necessity during a power failure.

Space heaters that burn gas or oil (heating oil or kerosene) are relatively large units, like wood stoves. They must be vented to the outdoors through an adequate pipe or chimney, and they must be connected to gas or oil supplies. Therefore, these heaters are more often installed permanently than as emergency standbys. Gas and oil heaters require relatively scarce and expensive fuels, another drawback.

As emergency heat sources, **wood-burning stoves** (and fireplaces) have the advantage over oil-, gas- and coal-burning heaters that they are versatile in the fuels they can use. If a power failure catches you without enough dry cordwood you can make do with less desirable fuels, which many country dwellers can get for themselves without relying on deliveries or specialist suppliers. Among these make-do fuels for wood-burners are: scrap lumber; bundles of sticks or twigs tightly bound into faggots; tightly rolled newspapers (a patented newspaper roller is on the market especially intended for making "logs" out of old papers) ; and any of several kinds of wood—notably sumac—that burn almost as readily wet and green as they do well dried. In any circumstances, however, avoid burning in stoves or fireplaces charcoal "briquets" intended for outdoor barbecue, or chemically-treated artificial "fireplace logs." Briquets give off highly toxic gases as they burn, and artificial logs burn very hot; they can break up and burn out of control. Finally, never burn trash in a wood stove.

If you use a wood stove regularly in your house you have standby heat immediately available in case of power failures. If you don't regularly burn wood, and live in an area that is subject to power failures, you might consider getting a small inexpensive wood (or coal) stove that can be kept out of the way and installed quickly when it is needed. This arrangement is not usually practicable for gas and oil heaters, which need to be connected to supply tanks.

As explained in Chapter 3, an adequate chimney must be provided for safe and efficient use of any stove. This applies as well to standby stoves that are hooked up for use only occasionally, in power failures and other emergencies, as it does to permanent installations. Even in a temporary installation, simply running the stovepipe out a window or passing it through a wall without insulating the passage is never safe. A convenient arrangement for small standby stoves is to set them in a fireplace with the stove-pipe running up into the chimney.

Smaller heaters are manufactured which do not need pipes or chimneys and which carry their own fuel supplies. These usually burn propane or kerosene. Propane models may run for about twelve hours on a tank of fuel; kerosene models run for about twice that long, but are more expensive (around $75 versus $25). The little heaters can be set up anywhere, but they may not produce enough heat to keep an average-sized room comfortably warm. They should not be used except in well ventilated spaces, and they should not be left unattended, for they can tip over and cause a fire.

Cooking in a power failure may or may not present a problem Obviously, the owner of an electric stove must have alternatives if he is to prepare hot food and water without electric service. On the other hand, some gas stoves are ignited electrically and some are not. Householders in power-failure-prone areas should buy gas stoves that do not depend on electricity. In any case, some standby source of heat for cooking—a portable camp stove, a charcoal grill (to be used outdoors only in both cases)—are a convenience in the snow-country house.

In a power failure, the electric pump or pumps that bring **water** into your house and distribute it through your water system stop. You need an electricity-independent way to procure water. If water gets to your house by simple gravity flow from its source your primary supply will not be interrupted by a power failure, though you will still need to bring the water from the house's gravity-flow-filled cistern or holding tank to wherever you need it. If your water is held in an open cistern in the cellar—a common arrangement in older houses—you can simply dip it out in buckets and carry it upstairs, or you can pump it upstairs with a hand pump. If your water is held in a tank you can tap the tank for the water you need.

If water is pumped into your house initially from its original source, you must husband carefully what you have when the power goes out, for your water supply will not renew itself until the power goes back on. In this case you may have to augment your water supply by carrying water from wells, streams or ponds, or, in winter, by melting snow. These are apt to be laborious expedients at best (it takes a great lot of snow to melt into a very little water, you will find), and may yield unpotable water. In a prolonged power failure you may need to use water which you have drawn off your house system for drinking and cooking, and snow water or water carried from outside for washing and flushing toilets (the latter is accomplished by dumping a bucket of water into the bowl).

It is wise to prepare for power failures by storing water in convenient places. In bad weather, if you think your power may go out, take time to fill your bathtubs and bathroom sinks, and any large vessels. This will assure you of at least some clean water ready to hand that you won't have to haul or pump.

Food stored in a **freezer** can spoil during a long power failure. The best way to protect your frozen foods from thawing and spoiling is to keep your freezer fully loaded at all times and to avoid opening the freezer while the power is out. Food in a fully loaded, closed freezer will usually keep for two days (food in a half-filled freezer will seldom keep longer than one day). A large freezer will stay cold without power longer than a small one; and a low chest-type freezer will stay cold longer than an upright model.

FROZEN PIPES

Another winter episode that tries the patience and tests the preparation of the snow-country householder is water pipes which freeze, putting some or all of a house's running water out of commission. In fact, frozen pipes can be more than a nuisance. Water expands when it freezes, and when it is confined in a pipe this expansion can crack the pipe. It will then leak when it thaws, perhaps causing damage.

Pipes are most apt to freeze on bitter cold mid-winter nights when the temperature drops below zero. Pipes do not, usually, freeze along their entire length; rather, they freeze at points that are exposed, especially where they pass near sills, corners or joints, or other uninsulated places.

The best way to deal with frozen pipes is to forestall them. Wrap fiberglass insulation or old newspapers around pipes, taping them in place, in segments of the pipes that can be expected to freeze—segments running along ceilings near outside walls, for instance.

If your pipes have frozen you must determine where they are frozen and then thaw them out to restore flow. **To locate the points where pipes have frozen,** know how the water flows through your entire system of pipes, then try turning on taps and valves at various junctures in the system. When you find a tap or valve that runs you have established that the freeze-up is "downstream" from that valve. You will probably find that your pipes freeze in the same places again and again (unless you insulate them) , where they are exposed to extremely cold air; and you will soon be able to return to these trouble spots whenever you have a freeze-up.

Once you have located the frozen points, you must **thaw the pipe out.** You can use a propane torch for this, fitted with a flame spreader to avoid concentrating the heat of the torch too narrowly. The torch will thaw your pipe quickly, but its use presents a fire hazard, especially where frozen pipes pass close to walls or woodwork. Another risk in using a torch is that it will heat pipes too quickly, causing them to crack. Better ways to thaw frozen pipes are to wrap them with electric heating cables (available at any hardware store) , to aim an electric space heater at them from close range, or to wrap them in rags and pour hot water over them. Obviously, the first two methods here cannot be used in a power failure. Obviously, too, none of the methods that apply heat directly to pipes should be used in houses with plastic pipes, which are used to save expense in some systems. If your pipes are plastic, thaw them with rags and hot water.

One precaution you can take against pipes freezing is to **leave**

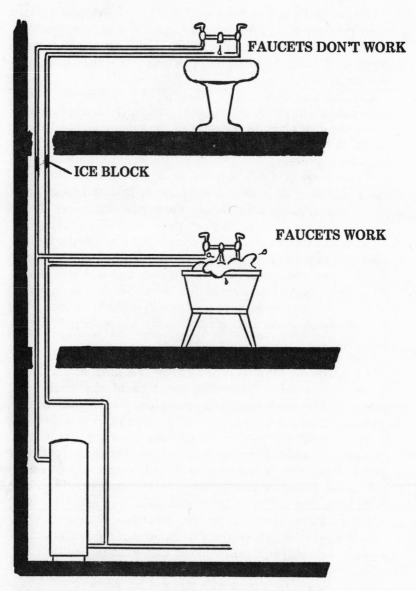

Finding water pipe freeze-up by checking faucets.

your taps running, at a trickle, whenever you expect extreme cold. The idea is to prevent freezing by keeping water moving through your system. If your house has plumbing on more than one floor, it shouldn't be necessary to leave taps open except on the first floor.

SNOWBOUND

During and after severe blizzards or prolonged snowstorms it can happen that houses are isolated for a time: people can't get to them or away from them—at least not on wheels—until roads are cleared. Unless your house is very remote it will seldom happen that you are isolated, or snowbound, for more than a couple of hours after a storm. Nevertheless it is advisable to make a few simple preparations against being snowbound.

If you have neighboring houses visible or within earshot you may want to establish a prearranged system of **distress signals** with your neighbors for occasions when telephone service is interrupted. Lights or flags make good sight signals, and bells, whistles or battery-operated horns (for use on boats) make good sound signals.

Have a two-day supply of **canned or dried foods** on hand to tide you over until you can get out, including food for any pets or livestock. Don't rely on frozen foods. As explained earlier, you want to avoid opening your freezer if your electric power goes off. If you or anyone in your house is on **medication** of any kind, make sure you have ample reserves of whatever the medication is. Keep a complete **first aid kit,** and know how to use it.

It may happen that fuel suppliers cannot reach your house when it is snowbound. In these circumstances, if you still have electrical power, you may find **electric space heaters** useful for emergency heat. Electric heaters are available in many sizes and types. Some expensive models, in which a liquid (usually oil) is heated electrically and circulated through a sealed radiator, are claimed to use less electricity to produce a given amount of heat than the more common models in which a metal radiant element is heated. Whatever kind of heater you buy, make sure it has a thermostatic control, and make sure that the radiant element is protected by a grill from coming into contact with combustible materials. Electric heaters should be used with care, and should not be left unattended: tipped over, they can start fires. In buying an electric heater, as with any other electrical appliance, check that its safety has been approved by the Underwriters Laboratory (UL).

Following any snowstorm a certain amount of **snow shoveling** is usually necessary to open up doorways, paths, driveways. It is important to remember that shoveling snow is hard work which can put unexpected and dangerous stresses on your heart and back if you go into it more vigorously than you should. Take steps to cut down the amount of effort involved in snow shoveling. If possible, shovel as soon as the snow has stopped falling: new snow is usually lighter and therefore easier to move than snow that has lain for a time. Use a shovel with a long handle to reduce stooping, and take small, partial shovel-fulls of snow. Rest frequently.

PREPARING YOUR HOUSE TO BE EMPTY

If your snow-country house is a vacation home which stands empty for long periods of time in the cold season, or if you make regular winter trips which take you away from home for more than a couple of days at a time, you must take steps to protect your empty house.

Unless you expect to leave your central heating system on while you are away—to keep your house from getting too cold—you can count on the inside temperature's dropping below freezing in the winter. You must shut off your water pump (if any), and **drain the water** from cisterns, tanks and pipes before you leave the house to stand empty; for if the water in your pipes freezes it may crack the pipes, and can flood your house when it thaws. At points in your house water system from which water does not easily drain (toilet bowls and pipe traps) you can add antifreeze, alcohol or kerosene—in small quantities so as not to damage your septic system. Don't forget to drain boilers and radiator pipes if your house has hot water or steam central heat.

Remove wood ashes from any stoves or furnaces before you leave your house. An empty, unheated house will grow damp, and damp wood ashes can produce caustic lye which will corrode the stove's cast iron or steel.

If you elect to leave your central heating on while your house is empty to keep undrained pipes from freezing, you will have to keep up your house electric service, too. But you may do better to drain your pipes and **turn off your electricity.** A prolonged power failure while your house is empty can result in frozen pipes even if you have intended to continue electric service, and a partial power failure ("brown out") can damage the motors in some electric appliances if you aren't around to turn the appliances off or unplug them. Frozen food left in the freezer of an empty house can, of course, spoil during a power failure. When you re-open your house after a long absence (during which your electric power was on—you think) check your frozen foods carefully.

Canned and preserved foods and liquids can freeze in an unheated house. Freezing will spoil food and drink in many cases, and frozen canned goods can burst their containers and make a mess. Give canned goods away before you leave, or take them with you, or put them in an area (like a cellar or root cellar) where you are sure the temperature will not go below freezing even in the absence of heat.

Animals can get into an empty house and wreak havoc with stored foods, upholstery, books and other goods. Mice and squirrels are the main offenders. Set out poison for them before you leave your house, but make a note of where you put the poison so you can retrieve it on returning to your house and remove its threat to children and pets. Food which might attract animal pests (grains, beans, cereal, pet foods) should be stored in tight glass or metal containers. Be especially careful to **store wooden kitchen matches tightly:** mice like to gnaw the matches' heads, which can cause them to ignite.

Apart from guarding against freezing damage by draining water pipes, the most important step you can take to safeguard your empty house is to arrange for someone to visit it regularly and check for incipient problems like ice dams, leaks, or missing shingles on the roof, broken windows and loose shutters. Ideally you will have a friend move into your house while you are away. As an alternative to this, you may be able to arrange for a neighbor, or the local police or sheriff's office, to keep a careful eye on your house. They must be able to reach you, obviously, to keep you posted on conditions at your house.

An unofficial caretaker or watchman—live-in or otherwise—is also the best provision you can make to protect your house against thieves. Burglary is big business in the country, where relatively isolated houses are especially vulnerable. While you are away, make sure an in-sight neighbor—if you have one—knows that anybody seen removing articles from your house is to be reported to police immediately. If you have no in-sight neighbors, you can at least ask neighbors to be alert to unauthorized persons on your premises.

As much as you can, make your house look occupied, even if it isn't. Get time-set switches for certain lamps (and have your neighbor or unofficial caretaker change the switch settings often). You might arrange to have your driveway plowed even if you won't be there to use it; a snowed-in driveway is a fairly sure sign of an untended house. Get a neighbor to pick up your mail every day. If your local paper suggests they would like to print an announcement that you are leaving on a world cruise, discourage them.

Thieves have favorite items: antique furniture, televisions, radios and all other electrical appliances, binoculars and other optical instruments, cameras, firearms, liquor—even firewood. Make certain when you leave your house empty that none of these items is in plain view from any of your windows. Make a list of the valuables in your house, including any identifying numbers, and file it with your insurance company.

A burglar alarm system for your house may help protect against break-ins. If your house is isolated, however, a conventional system—which sounds an alarm when it is activated—may not add any protection (for there won't be anybody around to hear the alarm).

"Silent" alarms, which alert the nearest police station when activated, are usually quite expensive, and may not be worthwhile if the local station is far from your house. Your state or local police or sheriff might be able to advise you on the real value of burglar alarms in your area.

5 Cars in Winter

ANYONE WHO HAS attempted to push a loaded wheelbarrow through a two-inch snow cover soon learns that the wheel and snow are basically incompatible. Snow, in this day and age, may be an ally recreationally: as far as transportation is concerned, snow is the enemy.

It was not always so. In the Revolutionary War, General Knox dragged artillery on sledges over the frozen landscape from Ticonderoga two hundred miles to Dorchester Heights to persuade British General Howe to evacuate Boston. Up until the present century, roads were rolled in the winter, rather than plowed, to accommodate vehicles with runners. But with the advent of the car, people expect they should be able to wheel their vehicles as readily in the winter as the summer, and the good winter road becomes the bare winter road.

Today a vast engineering and industrial complex has developed around the technology of snow and ice control. In Vermont, as an example, the cost of maintaining the state roads in wintertime has increased tenfold since 1931, from $72 per mile to $766. Snow removal cost metropolitan Montreal some $20 million in 1969.

And snow removal is just a start. A 1968 study made by the Society of Automotive Engineers calculated that rust destruction to private automobiles costs the owners about $100 per year. (A *Popular Mechanics* magazine story in 1975 puts the figure at $150 —"and more in areas where cars rust quickly.") Less obvious are the hidden but still huge costs involved with salt poisoning. Throughout the snowbelt coarse rock salt is dumped on roads by the millions of tons each winter to melt off ice and snow. No one knows the ultimate effect of salt in such quantities on vegetation and groundwater supplies, but there is evidence that salt poisoning is a crisis in the making in some areas. Nevertheless, the simple fact is that wheeled vehicles only operate efficiently on bare pavements, or at least a hard surface, and that salt, whatever its drawbacks, is going to be with us until something else is found that clears roads as well at a comparable or lower price. And driving in winter in snow country will remain slow, uncertain, usually unpleasant and often hazardous unless and until we can develop a national comprehensive passenger transportation system, and learn to live with winter comfortably instead of trying to fight it to a standstill.

This happy state of affairs is not immediately at hand nationally. But an individual approach to a more relaxed attitude towards snow and the automobile can begin right now in the driveway by applying what I call "The Philosophical Approach to Snow Removal." This approach reminds us that the Lord sent the snow and the Lord can take it away. The point is to quit shoveling snow around at great expense in time, energy, and money. During the winter of 1975–76 I practically eliminated removing snow from my driveway: instead I packed it down with snowshoes, rolled it down with the automobile—in short, turned it into a real, old-time winter road surface. I found that almost any snowfall (unless unusually wet and heavy) of up to four inches could be handled this way. For heavier accumulations I use (reluctantly) my driveway snow blower.

Getting around successfully really starts with the correct location and orientation of roads, houses, garages. You may be stuck with what you have. But if you do have the chance to build, and to determine the site of the building, put your garage near the road, and the house near by: a view is fine, but one pays for it in the high cost of plowing a long driveway.

Some people have compromised by building a garage near the road and then only shoveling a narrow pathway through the snow up to the house. This may be a poor solution, for it sacrifices the convenience of having ready accessibility to the house for deliveries of food and fuel; not to mention the hazard of fire equipment or ambulances not being able to get close to the house in case of a winter emergency.

Where winds and drifting snow are a problem, engineers, architects, town planners and others have sometimes made special efforts to lay out buildings and roads in such a way as to minimize their effect. For example, we know that if a road surface is raised a few feet above the surrounding land surface, wind will usually keep it clear of snow. Buildings can often be placed with regard to prevailing wind and each other in such a way that snow is deposited where it will do the least damage. In addition, across the snowbelt, snow fencing and shelterbelt planting, judiciously placed, have played an important role in snow management.

CARS FOR SNOW COUNTRY

Since we rely almost entirely today on the private automobile for getting around from place to place, we should choose one that starts well in cold weather and handles well on snow-covered and slippery roads. For anyone who lives in snow country, performance and handling in winter should be a prime, if not the first, consideration in selecting a car.

The most important single factor in any car's winter performance is **traction**: How well does it hold the road in ice and snow? How apt is it to slip, spin and skid? In turn, the most important factor in determining traction is weight: How is the total weight of the car distributed relative to the drive wheels? The higher the percentage of the car's total weight that is centered over the drive wheels, the better traction a car will have. In cars designed for maximum traction, weighting the drive wheels is usually achieved by locating the engine over them. Since the demise of the popular rear-engine Volkswagen "beetle," most cars having engines over drive wheels have their engines in front, and power the front wheels. Most such cars presently (1977) on sale in the United States are imported from Europe or Japan; American cars are generally designed for front engine, rear drive. On the principle that cars in which a relatively high percentage of total weight is located over the drive wheels are cars which have good traction, front-engine–front-drive makes are better winter cars than front-engine–rear-drive (American) makes. Many experienced drivers feel that imported, front-engine–front-drive cars corner better than other designs on slippery roads; though others feel that, overall, American cars are more maneuverable in all winter conditions.

Owners of American cars who feel that their car's traction would be improved by increased weight over the drive wheels often weight the rear by putting sand bags in the trunk or under the back seat when winter approaches. The sand supply is also useful in other ways: you can throw handfuls of sand under wheels spinning on ice to help your tires grip.

Although weight and weight distribution are the main factors in traction, other car features can increase traction. Among these features are transmission and tires.

Four wheel drive and "limited-slip differentials" are transmission features which are intended to aid traction in bad driving conditions. The best traction comes with four wheel drive. Normally four wheel drive vehicles are driven with power going to one pair of wheels only, for more economical operation. But for off-road use and for driving on slippery roads power distribution can be changed manually so all four wheels have power. On some of the newer four wheel drive makes, there is a special option which automatically shifts power to the off wheels when needed. Automatic, or "full-time" four wheel drive, which is expensive, is an advantage in driving over changing surfaces, for example on roads which are iced over in patches.

In vehicles equipped with "limited-slip differentials," only one pair of wheels is driven, but extra power is transferred to whichever drive wheel has traction when the *other* is slipping.

It used to be an article of faith among snow-country drivers that cars with **manual transmission** handle better in winter than cars with automatic transmission. Unquestionably manual transmission allows the driver to control the amount of power going to the drive wheels as he cannot with an automatic, and thus is frequently an advantage in winter driving. Drivers of manual transmission cars can also get their engines started by rolling them downhill or pushing them, as drivers of automatics cannot—another advantage in a climate where low temperatures may make any car hard to start. Nevertheless, many experts today feel that—with automatic transmission being offered in more and smaller cars—an experienced driver in an automatic-transmission car can get on as well in snow and ice as the driver of a manual transmission.

Tires are more important to successful, safe winter driving than any other piece of car equipment. Before the advent of radial tires (see below) most experienced drivers in snow country regarded deep-tread snow tires as a necessity. Most drivers mount snow tires only on the drive wheels of rear-drive cars, but some makers of cars, including front-engine–front-drive models, recommend snow tires on all four wheels as an aid to braking. According to the National Safety Council (NSC) snow tires give twenty-eight percent better traction than regular tires on glare ice at twenty-five degrees Fahrenheit (3.9 Celsius) —conditions which prevail all too often in the northern snowbelt. Many drivers who travel unpaved country roads much of the time leave snow tires on their drive wheels year around, for added traction in mud, loose gravel and dirt as well as in snow and ice. They switch to the newer of two sets of snow tires for use in winter.

Added traction can be achieved by setting steel studs into the treads of snow tires. Studs make more difference on hard ice than they do in deep snow, but in any case they are an important aid (218 percent improved traction over regular tires in the NSC tests with new studs; 183 percent with used studs). Disadvantages of studded snow tires are that studs wear away with use and cannot ordinarily be replaced; and studs are also noisy when tires equipped with them are rolling on dry pavement. Also, some states prohibit studded snow tires because they damage road surfaces.

"Sandpaper" re-treads, not as common now as they have been, are sometimes an alternative to modern snow tires. In these, an abrasive compound is mixed with the hot rubber when new tread is applied to a used tire casing. Sandpaper re-treads give good traction on glare ice, but wear down very quickly.

Tire chains, the most cumbersome way to improve your tires' traction, are also the most effective. Chains on the drive wheels improve ice traction 630 percent, according to the NSC tests, and improve traction on snow by 313 percent. For emergency use, short segments of chain link ("helper chains") which can be easily belted onto tires are available, and are far easier to put on than tire chains. Another variation on the conventional steel reinforced tire chain are non-metal chains which are being developed. Made of a new polyurethane compound, Elastomer, these chains can be left on tires indefinitely, their makers say. Cars equipped with Elastomer chains are supposed to be able to be driven on dry pavement without damage to tires and without undue vibration.

The winter tire situation has been complicated recently by the introduction of **radial tires.** A radial tire differs from a conven-

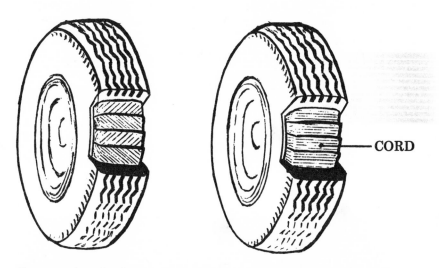

CORD

Cutaway view showing cord alignment of radial (right) and conventional, bias-ply tires.

tional, "bias-ply" tire in that the structural cord of a radial is arranged horizontally while the conventional tire's cord is laid on the bias or diagonal (see drawing). When a vehicle comes from the dealer equipped with radial tires in the normal, or summer, tread pattern, and the owner wishes to use winter tires, it is important that these be radial tires, too, because mixing radial with bias tires dangerously hampers the handling of automobiles: radials are not compatible with bias-ply or belted tires. People who live where the snow problem on roads is not too severe may find that radial tires with the summer tread have sufficient traction for their purposes in winter. It is a fact that radial tires give better traction that bias-ply tires both on bare and slippery pavements.

Some European tire manufacturers have recently announced a new type of rubber used in making radial tires that gives outstanding performance in cold weather and on slippery roads. This "high-hysteresis" rubber, so-called, is claimed to stay flexible at low temperatures so that its gripping power is superior in winter. These new rubber radials will make studded snow tires obsolete in a few years, their makers claim.

WINTER DRIVING ACCESSORIES

Light hardware with which every snow-country car should ideally be equipped includes the following:

Engine block heater
Shovel
Ice scraper/snow brush
Tow rope, chain, cable (20 foot)
Sand or traction mat
Tire chains
Battery booster cables
Cable ratchet come-along
    and extra cable or chain
Blanket

Flashlight
Towels or rags
Flares
Plastic windshield sheet
Winter windshield wiper blades

Engine block heaters, usually electric, are an advantage for drivers whose cars must stand for long periods (for example, over night) exposed to below-zero temperatures. An electric heating element warms the engine oil, engine coolant solution, or battery (cost: $6–$15). Use of any of these heaters requires access to house electric current. Propane and kerosene engine heaters are also available, and need no electricity, but they may constitute a fire hazard. If you don't have a heater, and you must leave your car out all night, nose it into a snowbank, or at least park it so the engine is out of the wind.

Come-alongs are cable hoists equipped with a ratchet stop. With one end hooked onto a stuck car's axle and the other anchored to a tree or post, a come-along enables you—working alone—to pull your car out of the ditch if you have gone off the road. Come-alongs are fairly expensive ($20 and up), but they make a good investment for anyone who does considerable winter driving, especially for travelers of rural back roads, which are apt to be slippery, to have deep ditches, and to be near trees to which a come-along can be conveniently attached.

Winter wiper blades (costing about $7 a pair) can be a real convenience. A rubber sheath, backing the blade itself, covers the metal parts of the windshield wiper assembly, preventing those parts from collecting ice and snow. If you replace winter wipers with regular blades in summer, there is no reason a pair should not last for several winters, at least.

The uses of most of the other accessories on the list above are self-evident. A plastic sheet secured over the windshield of a car which must stand over night can save a lot of ice-scraping the next morning. Traction mats are placed under spinning wheels of stuck cars to give them a purchase.

WINTERIZING

Cold weather, even snow, frequently arrives unexpectedly. If you use snow tires, they should be mounted before the time snow is expected in your area. Your car should also be winterized. Not only does winterizing help insure easier starts and more enjoyable, safer cold weather motoring, it will save gas, too. A tune-up—the first step in winterizing—can increase fuel economy by as much as fifteen percent, according to the American Automobile Association (AAA).

Normally, a tune-up consists of adjustment or replacement of spark plugs, as required; oil and oil filter change; air filter change; adjustment or replacement of points; new condenser, if needed; inspection of heat riser valve; carburetor adjustment; pollution control valve change, if required; tightening or replacing of belts; and replacement of ignition wires, if necessary.

For winter maintenance, also ask your mechanic to check anti-freeze solution. Most cars should have antifreeze changes at two-year intervals. Water hoses and the exhaust system should also be inspected for leaks, and battery charge and capacity should be checked. Ask the mechanic to clean battery terminals and grease them to guard against corrosion—a major inhibitor of engine starts, especially in cold weather. A fully charged battery has only sixty-five percent of its starting ability when the temperature drops to freezing, and at zero a battery has only half its power at seventy degrees. If your battery is three years old or more, you might consider replacing it with a new one before really cold weather sets in.

When you change your oil, make sure the weight, or viscosity, of the oil you have been using in summer will perform satis-factorily in winter. A multi-grade oil (like 10w–30 or 10w–40) works well in almost all engines and has a flowing consistency to meet varying temperatures. It may be necessary to change your car's oil more frequently in winter than in other seasons, for oil can become diluted with gasoline in cold weather, especially in cars driven in stop-and-go traffic. See your car owner's manual for recommendations on oil weight in different seasons.

Among car systems which require special attention in winter are **brakes.** Disc brake mechanisms are fully exposed to water, salt, sand and slush. See that brakes are checked and their mechanisms cleaned regularly in winter.

The AAA also recommends that the following be checked in preparation for winter: lights, turn signals and flashers; windshield wiper blades (periodic cleaning with a household cleaning fluid can sometimes renew the wiping power of dirty blades and extend their life) ; windshield cleaning fluid supply (the fluid must contain antifreeze) : heater and defroster.

Wind leaks at doors and windows are a major irritation in winter. Owners of older cars, particularly, should check weatherstripping and door and window fits for needed repairs.

One last word of advice on the subject of car maintenance: read your **owner's manual.** Owner's manuals, free to about eleven million Americans each year, may be the least read bestsellers ever published. The people who made your car are your first source of advice, through the manual, on how to keep it up. Emission control systems recently introduced, for example, have caused auto makers to revise start-up instructions for many new cars. Drivers who fail to follow these procedures may have difficulty starting their cars, especially on cold mornings.

WINTER DRIVING

Winter driving is something almost all of us who live in the northern snowbelt feel we know something about. But, according to the NSC's Committee on Winter Driving Hazards, we're not always right. This committee has been testing vehicles, equipment, and driving techniques under real winter driving conditions since 1939. Their tests make it clear that misconceptions and bad habits in the area of winter driving don't decrease as you go north.

The First Law of Winter Driving is: *Slow Down.* Low speed makes it less likely you will lose control of your car on slippery roads, and, if you do lose control, low speed will give you more time to get it back. Good winter drivers treat ice and snow with respect and gentleness—they start gently, stop gently, maneuver gently.

Avoid sudden pressure on brake or accelerator, and avoid sharp turnings of the wheel. In whatever conditions—slick ice, packed snow, loose snow—try not to break the tenuous bond between tire and road. If you accelerate too fast the drive wheels will spin and polish the surface under the tires. When you stop, do so gently, starting to brake far sooner than you normally would on clear pavement. In braking, if your wheels lock and you start to slide,

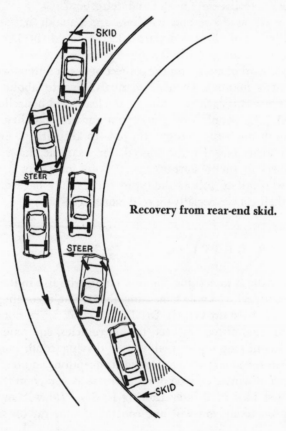

Recovery from rear-end skid.

pump the brakes—apply your foot on the pedal once or twice and release it once or twice a second in a repeating action. This lets the wheels turn and provides a measure of steering control, which you don't have if your wheels are locked. With locked wheels, your momentum will simply carry you in the direction you are already going.

If you go into a **rear-wheel skid,** steer in the direction of the skid, that is, if the back of the car slides to the left, turn the wheel to the left. And then, when the rear wheels start to line up with the front, straighten the wheel. Be prepared to counter-correct if your rear wheels swing over in the other direction. Sometimes it may help to disengage the clutch briefly at the moment a rear-end skid begins. A **front-wheel skid** is best managed by turning the front wheels in the direction of the skid momentarily, until the wheels can get a new grip on the road's surface. Above all, in any skid, use the brakes rarely if at all. Braking usually aggravates a bad situation.

Always be on the lookout for slippery patches on an otherwise bare road—as for example, in shadowy areas where the sun cannot reach. Sometimes the surface of a bridge or overpass will remain icy when the roads around it are clear: do any braking you must do before you are on it.

"Wet" ice, like that produced by freezing rain, presents possibly the most treacherous of all driving conditions. Avoid driving in freezing rain if at all possible. Another especially hazardous condition arises when two inches or so of snow have fallen, with light winds blowing and temperature in the low twenties. This snow may have melted on roads from pressure of tires, re-frozen, and then been covered by new snow. The innocent-looking road can be lethal.

If you have any doubts about the state of slipperiness of the road, find a place where there is no traffic, slow down, and firmly press your brake pedal: you will be able to tell what the condition of the road is by how fast you slow down and whether you start to slide.

The "bare road policy" adopted by most highway departments today has the effect of encouraging drivers to travel at high speeds

—even when ice, unseen until the last minute, may be encountered around a corner. It is doubly important in the winter to practice the fine art of **defensive driving**, especially on highways during storms. Be alert for other cars losing control. Double or triple the normal safe interval between yourself and other cars. Remember that even at twenty miles per hour it can take you ten times more distance to stop on ice than on dry pavement.

In town and city driving, motorists can avoid unnecessary braking by down-shifting to a lower gear as traffic slows. This will slow the car without the need for heavy braking.

You will be safer in winter driving with your **headlights** on, long after official sunrise and long before official dusk, the most accident-prone times of the day. Always keep your lights on in snow or rain. If your alternator is working properly, there is no danger of "draining your battery," as some apparently fear. If traffic comes to a stop when visibility is poor, turn on your four-way flasher lights to be sure vehicles following will see you and have time to stop.

When you can—that is, when there are no other cars on the road—drive in the middle of **unpaved country roads** when the going is bad in winter. The surface is usually harder in the middle, there is more room to maneuver if something jumps out in your path, and your right-hand wheels won't be as likely to get caught in soft snow or a hidden ditch. (If your wheels start to pull to the right in these circumstances, try *gradually* to ease back to the left onto a harder surface; a sharp turn here might actually result in the front wheels getting drawn deeper over to the side.) Obviously driving in the middle, or close to the middle, of the road can only be attempted when the way is clear ahead and you have ample warning of the approach of oncoming traffic.

Meeting and passing other vehicles on icy country roads should be done at a crawl. When meeting an oncoming car on a hill, give the driver of the car headed uphill the right of way. At night, turn off your headlights briefly—leaving your parking lights on—to give the oncoming driver every advantage in getting up the hill.

On blind corners of back roads, or when approaching hidden

side roads, be liberal in the use of your horn: a few blasts in the right places could give helpful warning to other drivers.

Sometimes on rainy or slushy roads a wedge of water builds up between the tire and the road surface ("hydroplaning") which can cause loss of traction and steering control. To forestall hydroplaning accidents, slow down, watch out for standing water, replace worn tires and keep tires properly inflated, and keep a safe following distance from other vehicles.

Hill climbing on icy country roads is a special challenge. The most important point is to get up speed—always being sure there is no traffic in the area—and keep it up. Before starting up the hill, choose the highest gear ratio which will allow you to get to the top without shifting again. Losing headway on a long, slippery hill is fatal for your chances of getting to the top. On the other hand, high speeds in these conditions can be fatal, literally: thirty to thirty-five miles per hour should be enough to give you the momentum you need. Don't try to barrel around a group of cars stuck on a hill. Pull over and try to find another way around the hill.

If you bog down in **soft snow,** use second gear and attempt to crawl forward slowly. If tires spin, shift to a lower gear and concentrate on keeping the car in forward motion. Letting several pounds of air pressure out of the rear tires may improve traction slightly.

If you get stuck in snow you can sometimes "rock" your car out of its position. Shovel away snow from your car front and rear and, as much as possible, underneath. Make sure you have cleared a path for your wheels in the direction you want to roll to get out. Shift back and forth between reverse and the nearest forward speed in your car's gear sequence, accelerating gently as the gears engage. This should have the effect of rocking your car in the elongating ruts your wheels have dug for themselves in the snow, allowing you to roll out of them. If you have helpers, let them push you as you accelerate. Should the car remain stuck after several minutes of rocking, have it towed to avoid overheating and possible damage to the transmission (or try to pull it out with your come-along) . **If you get stuck** and can't get out, police advise staying with your car unless you know for sure that you are near help and that you are capable of hiking through the snow to reach it. If there is a chance you might be marooned for some time, treat it as a serious situation. If you have enough gas, you can run the engine. But *don't* do so if you have any doubts about the condition of your exhaust system. And always be on the alert for exhaust gases. Make sure the car's tailpipe is not blocked before you start the engine, and if heavy snow is drifting someone should get out periodically to clear it (and to be sure that one door can always be opened) . Try to keep parking and dome lights on as a beacon for rescuers. And open the window a little on the side away from the wind. (Incidentally, run the engine at a speed equivalent to about thirty miles per hour, rather than letting it idle: this actually uses less gas, and keeps the battery charged.)

Visibility is one of the major problems in winter driving. All modern cars are equipped with windshield defrosters, and many have optional rear-window defrosters as well (these last are an important advantage for all who drive in winter). Defrosters, however, can be a mixed blessing in some circumstances. In a snowstorm when temperatures are around freezing, a defroster blowing hot air onto your windshield can warm the glass, causing snow to stick to windshield and wiper blades. Driving in such a storm, you do well to turn off your car heater and blow *cold* air on the windshield. This will help the glass shed the snow. Don't operate your windshield wipers until you are sure your windshield has cooled.

If you must travel by night in a heavy snowstorm or blizzard you will find the driving snowflakes can reduce visibility to near zero. Pull over and wait out the storm if you can. If you must continue, put down your sun visor, turn headlights on low beam, switch on your four-way flashers and reduce speed to fifteen to twenty miles per hour. If you have a passenger let him or her keep watch ahead. You do not watch the road ahead, but concentrate on the side of the road to your right, where the guard rail should be clearly visible to guide you.

TROUBLE

Winter in snow country presents an assortment of small crises for the motorist, including locks and hand brakes which freeze, cars which won't start, and gas lines which ice up.

Just getting into your car can be a problem if the door lock is frozen. Heat the car key with a match to thaw out a frozen lock. A lock lubricant should prevent future freeze-ups. Don't set your emergency brake on an extremely cold night: it could freeze in place.

Easy cold-weather starts eliminate unnecessary wear and tear on the driver as well as on the car. To start most cars on a cold morning, activate the automatic choke by pressing the gas pedal slowly to the floor with the ignition key in the On position. Then release the accelerator completely and turn the key to Start. If the engine floods after several tries, hold the gas pedal to the floor while turning the key. If the weather is extremely cold (*minus* forty degrees Fahrenheit) try bringing the battery indoors at night.

An unstartable car may require a hook-up to a booster battery with jumper cables. Most motorists don't know how to use jumper cables properly, and if improperly used cables can be dangerous. Here's the correct procedure for **jump-starting a car.**

1. Make sure the stalled car and the car with the booster battery are not touching.
2. Make sure the two batteries have the same voltage.
3. Turn off battery-operated accessories such as headlights, radio and heater, to eliminate unnecessary power drains.
4. Set the emergency brake on the car you want to jump-start and shift transmission into Park (automatic) or Neutral.
5. With the engines of both cars off, connect one end of either of the two booster cables to the Positive terminal of the booster battery. Then attach the other end of the same cable to the Positive terminal of the weak battery. The Positive

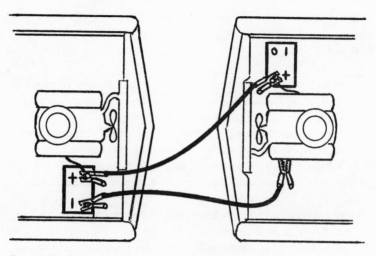

Correct hook-up of cables for jump-starting car on right.

terminal on most batteries will be marked with a plus (+) sign and will have a cable leading from it to the starter.

6. Connect one end of the second cable to the Negative (−) terminal of the booster battery, and the other end of the same cable to a ground connection on your disabled vehicle, such as a bolt on the engine block. It is best not to connect to the Negative terminal of the battery to be boosted.

Now the jumper cables are connected properly. Start the engine of the car with the booster battery and turn your own car's ignition to Start. If this first attempt fails to start your car, let the boosting car idle for a few minutes, charging your battery, then try to start your car again. Once your engine is running at normal idle speed, simply reverse the hook-up procedure to remove the cables. Begin with step six, removing the ground connection, and move back through step five. Boosting a weak or dead battery may seem easy enough. However, experts warn that because air conditioning and other power accessories demand higher-capacity batteries on newer cars, there is good reason to use caution. Never use booster cables on fuel-injected cars, or on cars with so-called "computerized ignition"; doing so may damage delicate equipment, necessitating costly repairs.

Condensation moisture in your fuel line may freeze in winter, cutting off the supply of gasoline to your engine. Pour a can of gas-line antifreeze into your tank and wait a few minutes for the ice block to free. To keep moisture from forming in your gas line in the first place, it helps to keep your gas tank filled in winter. Another advantage of a full tank is that it adds weight to your car, improving traction: twelve gallons of gasoline weigh about 100 pounds.

A final tip—never more obvious than when you have come to grief through its not having occurred to you: If you keep your car in a garage overnight it is good practice to back it into the garage during winter. Should there be heavy snow during the night it will be easier to get through the new snow head first, and the car's engine compartment will be more accessible to jumper cables from another car if the car is heading out.

Any discussion of "tips for winter driving" such as I have given above is bound to strike terror into the hearts of unconfident drivers everywhere. It seems impossible to remember the right thing to do in a given set of circumstances—circumstances which will seldom allow time to ponder alternatives. Indeed, winter driving is difficult: it must be learned and re-learned each winter by everyone. But in time you do get the hang of it, and each winter (after an initial rusty period shared by even experienced drivers) you become more proficient. As you are getting the hang of winter driving, and after you have it, the most important rule to remember is the First Law: *Go Slow*. The rest will come.

6

Dressing for Winter

CLOTHES MAKE THE MAN, they say, but not just any clothes will make the comfortable man—at least not in winter. Dressing comfortably in winter takes some understanding of the way our bodies use—and lose—heat.

To operate efficiently, the body must maintain a temperature of close to 98.6 degrees Fahrenheit (37.4 degrees Celsius). At rest, the body produces heat to sustain this temperature through normal digestive processes; physical activity also contributes body heat. We wear clothing to conserve body heat and prevent its loss to the surrounding air.

Body heat is dissipated in three ways: by convection, radiation and evaporation. Air in motion around the body carries off body heat by convection, an effect which is increased by outdoor conditions, like wind. At the same time the body, like a furnace, loses heat by simple radiation: in effect it is heating the space around it (this is why we "huddle together for warmth" in a cold place). Finally, if the body—through exercise, perhaps—builds up more heat than can be dissipated by convection and radiation, perspiration evaporates to cool it further. The effect of evaporation, too, is increased by wind, for air in motion evaporates moisture more efficiently than still air, and therefore carries off more heat.

In general, winter clothing is intended to prevent loss of body heat by convection, radiation and evaporation. In the case of evaporation, however, a delicate balance must be kept to achieve comfort in winter; for, if perspiration is prevented too effectively from evaporating, the damp body and clothes can become chilled.

WINTER CLOTHING

In the chapters of this book on weatherproofing and home heating the function and importance of insulation have been obvious. Clothing is body insulation. Just as house insulation works by holding still, "dead" air in pockets between the heated area of the house and the outdoors, so warm winter dress should hold air warmed by body heat next to the skin. This is best accomplished by wearing clothes in several layers. **Layering** is the key to winter dress: rather than putting on a single heavy coat over your indoor clothes when you go outside in cold weather, put on a sweater or vest over your shirt before you get into the coat. This gives you three or four layers of clothing, and between these layers warm air insulates your body. It is the layers of air that keep you warm, not (or not so much) the material of the clothes themselves. The principle that winter clothing functions as insulation has led to the development of the present variety of construction types in clothes of whatever materials. "Waffle" weaves, fishnet, quilted clothing, bonded and pile fabrics, and others are all intended to trap and hold air, exploiting the idea that clothing must insulate.

KEEPING WARM INDOORS

Before the continuing energy crisis required snow-country householders to mind their thermostat settings, indoor winter clothing was often dictated by convenience or fashion rather than by warmth. Today, however, as the thermostat goes down the clothing goes on: sweaters cost less than auxiliary space heaters, and they're easier to install. Dressing warmly inside allows you to be comfortable at a lower temperature, and, as mentioned in Chapter 3, even a slight lowering of temperature can result in a measurable fuel savings.

Obviously there are no hard rules for comfortable indoor winter clothing. Some houses are much warmer than others, and their occupants dress accordingly. Nevertheless, even if your house is a warm one, reflect: wherever your thermostat is set, if people in your house aren't wearing warm clothes inside, it could be set lower if they were.

As a general rule, for outdoor wear as well as indoor, "natural" materials—cotton, wool, and mixture weaves like Viyella—are warmer than synthetics. The main warmth advantage of natural over synthetic materials is that the former allow free evaporation of perspiration, while the latter are more apt to retain it. Indoors, wear lightweight wool rather than cooler synthetics. Wear long-sleeved sweaters (someone has calculated that putting on a sweater can have the effect of a 3.7 degree increase in thermostat setting) , and, for the ladies, long skirts in the evening. Long underwear or tights worn in the house are good for those willing or wanting to push thermostats down even lower.

In choosing **footwear**—for indoor use as well as out—remember that your feet are at the end of your body's warm blood line and therefore they feel undue loss of body heat early and acutely. Loose fitting, warm footwear does a lot to keep body heat evenly distributed and comfort up. Sneakers with wool socks or tights are usually ample for indoor daytime wear. Lined slippers, slipper socks, and down-filled booties are excellent for sitting. Booties filled with synthetic fiber instead of down can even serve for occasional quick trips outdoors, for their insulating material stands wetting better than down does.

Nightwear has its own requirements for winter comfort. The general consensus now is that it is more economical to turn your electric blanket up and your thermostat down than the other way around: it's easier to heat a bed than a house. Again, though, clothing can give you alternatives to heating. Flannel or brushed cotton nightwear—or thermal underwear—can help keep you warm in a cold sleeping room. Silk is also warm, though silk garments are rarely cut for warmth. Remember that your head is an especially easy place for heat to escape from because blood vessels travel near the surface there over your temples and skull. A nightcap can therefore help keep you warm out of all proportion to its size.

Apart from nightclothes, down-filled comforters, warmer than any blanket, can substitute for electric blankets in some situations. They are quite expensive, so borrow one to try out, if you can, before you buy. Hot water bottles, warm bricks wrapped in a towel, or even fireplace coals in an old brass warming pan, can go a long way toward making your bed a more hospitable place on a cold night.

If you must have an electric bed-heater of some kind, consider an electric mattress pad in preference to an electric blanket. The mattress pads cost half what the blankets cost, and are more efficient, since their heat comes from under you and warms your body, the bed and the blankets before escaping, while the blankets heat the air above the bed as well as they heat the sleeper.

Relying on warm clothing to help effect fuel savings will be harder in some households than in others. As pointed out in Chapter 3, homes with live-in or visiting older people or invalids may require relatively high thermostat settings. Houses with very young children—crawlers and floor sitters—pose still another problem (thick, warm rugs will help there). Extra clothes, high spirits and a little imagination may not always win the day, but they will always help.

It is also important to "think warm." Keep your house bright with warm colors—bright throws and furniture scarfs, dried flowers, pictures, all add cheer. Dress brightly, too: colorful clothing will warm the spirits of those around you as they warm their wearer.

OUTDOOR CLOTHING

Besides the principle of wearing clothes in layers to retain warmth, the most important factor determining outdoor winter clothing is **activity:** what will you be doing? The needs of those doing strenuous work or play (snow shoveling, wood cutting, skiing) are different from the needs of more sedentary outdoorsmen (ice fishermen, some hunters, spectators). The more you do, the less you wear.

The layering principle takes account of the varying clothing needs imposed by different levels of activity. If you wear several layers of clothing, you can remove layers during periods of activity and replace them when you are at rest. This gives you flexibility: you can adjust your insulation to meet changing requirements.

The first clothing layer for anyone who lives in snow country should be **long underwear.** Some people still prefer old-fashioned woolen underwear (now made with soft "itch-free wool" and with a small admixture of nylon yarn for strength). Especially when combined with a lighter suit of underwear (usually fishnet) woolen underwear is exceptionally warm. It is expensive, however, costing almost twice as much as cotton "thermal" underwear.

"Thermal" underwear, so-called, made of cotton, is the most popular underwear for all-purpose cold weather wear. Thermal underwear comes in three types: knit, fishnet and insulated. Knit is the most commonly used of the three, and it is usually sold in either a smooth or "waffle" weave pattern. Fishnet is open net fabric, and is usually worn under regular underclothing. The air spaces in the net make the fabric warm and light. Insulated, or two-ply, thermal underwear is the warmest of the three. It consists of an insulating space between two layers of cotton knit.

Down-filled underwear is also available, but it has a drawback (in addition to its cost, which is high). Though down has exceptional insulating qualities, it is bulky, and further, it does not allow for easy evaporation. Once it gets wet, either from perspiration or from an external wetting, down loses much of its insulating quality, and, at the same time, it holds moisture.

Thermalactyl is a fairly new synthetic material for heavy-duty winter underwear. It is claimed both to retain and to reflect back body heat, thus helping to promote circulation in the tiny blood vessels just below the skin. Unlike most other synthetics, Thermalactyl also allows for free evaporation, its proponents say. Underwear of Thermalactyl might well be worth a try if you plan prolonged exposure to severe cold.

Over whatever underwear you have, you will want to add a cotton shirt or T-shirt, turtleneck or light woolen shirt. Your third layer could be a sweater, sweatshirt, or lined vest. As else-where, materials should be natural in preference to synthetic: cotton or wool, or Viyella.

Below the belt it is less convenient to add and remove clothing layers at will. Trousers, therefore, should be warm and sturdy: wool or gabardine for preference. (Cotton jeans do not make good winter wear, for when they are wet they are a very poor insulator, conducting heat away from the body 240 times faster than the same material does when it is dry.) For hard outdoor wear, a pair of heavy woolen breeches, with reinforced knees is a good buy. They should be cut to fit loosely at thighs and more tightly over the lower legs, allowing boots to fit well up over the ankles. Warm-up pants are worth considering if you are active sporad-ically in severe cold. They are often slit at the outsides down each leg, for ventilation, and so you can quickly zip them on over your pants when you stop what you are doing.

Your final clothing layer will be a coat of some kind: wool or gabardine jacket or heavy parka. For warmer winter days, or for active work near your house, a light-weight, wind-resistant zip-pered jacket—worn over other clothing—will keep you warm and unencumbered. A light lining or fiber-fill stuffing will make such a jacket even warmer.

For cold weather wear away from shelter, and for less vigorous outdoor activities, you will need a parka. **Parkas** are rarely wanted during normal activity, except perhaps in extreme cold, but they are very welcome between periods of exercise. They come in a great variety of styles and materials, all comparatively expensive. Wool parkas are warm, but are apt to be bulky and heavy compared to newer down- and synthetic-fiber-filled parkas. Since you won't usually wear a parka while you are being es-pecially active, the superior evaporation qualities of wool over synthetic fabrics may not be an important consideration in your choice of a parka. However, wet wool parkas will dry out better and faster than synthetic ones, which may be an advantage.

Fiber- and down-filled parkas are more expensive than woolen parkas, but they both have superior insulating qualities, and are much lighter than wool. The best **down** comes from geese, whose feathers have a high lanolin content, making their down relatively moisture resistant. In the United States, however, most clothing manufacture uses the more abundant and less expensive duck down. To judge the quality of down in a garment, squeeze it through the fabric; it should be soft and springy, free of quills or spines.

Good down-filled parkas are very expensive, though recently several mail-order houses have begun to offer down-filled parka kits (sew it yourself) at about one-half the price of the finished product. People who have used these kits report satisfaction: kit-built products are durable, and kit directions generally easy to follow, though some makers have trouble sewing thick, layered materials on small sewing machines. (See the end of this chapter for names and addresses of two down kit suppliers.)

Synthetic polyester fibers are used today as insulating fill for garments where down would have been used a few years ago. The polyester fibers are not as well suited for clothing insulation as down: they are heavier than down, not as long-lasting, and more easily compressed. Polyester fibers stand wetting better than down, though, and are usually less expensive. These fibers are sold under a number of brand names. Two of the best and most popular are Hollofil II (manufactured by DuPont) and Polar-Guard (by Celanese).

A new material for the outside shell of down- and fiber-filled parkas is a special fabric called "Gore–Tex." Gore–Tex was originally developed by medical researchers for use as artificial blood vessels. Laminated and produced as more conventional material, it is permeable to vapor but not to liquid or air. Used as a shell around a down filling, Gore–Tex may prove to be the outer-wear fabric of the future: for it will prevent moisture from reaching the down interior of garments made from it while allowing evaporation to take place. Ask your favorite sporting goods dealer about Gore–Tex. It is very expensive, but if you are in the market for a parka, costly new materials are worth investigating. (All major winter clothing items are expensive, and all are important to your comfort: it's worth your time to shop around.)

Animal fur also makes excellent parka material: it is warm and it permits evaporation. Most fur (including the new artificial furs) is far too expensive for use in practical outdoor clothing, however.

An unusual parka fill material, for those determined both to try something new and to economize, is milkweed floss—the silky, white fluff attached to seeds of the milkweed plant (which is common over much of the northern United States). It has been estimated that milkweed floss is six times lighter than wool, and equally warm. Gathering floss to stuff a parka would be laborious, and the parka would require care in washing, but the experiment might prove to be worth the time and effort.

Among other variable features of coats and parkas that are worth the buyer's attention are **color, fastenings, hoods and pockets.** In selecting winter clothing remember the contribution to heat absorption made by color: dark material absorbs more heat than light material. Zippers and Velcro are better fastenings for winter clothing than buttons. Buttoned coats let in drafts, as coats closed with zippers and Velcro do not. Also, buttons can be awkward for stiff fingers to work in the cold; zippers (especially when equipped with a short rawhide thong) and Velcro are more convenient. Drawstrings at the hip and waist of parkas help close out the wind. Hoods, with drawstrings around the face and chin, help keep your head and ears warm, but cut peripheral vision and impair hearing. Hand pockets are a real comfort, but they should have zipper closings, as should any other winter outer wear pockets. There is little warmth in putting your hand into a snow-soaked pocket.

All winter wear should be loose fitting, but outside garments must also keep out the wind, especially at the ankles, wrists and neck. These points, where the blood circulates close to the body surface, are especially subject to heat loss. In well-made outer wear, tight-knit cuffs and ankle bands protect vulnerable spots.

Neck scarfs add a welcome bit of color to winter life, and keep the neck and chest warm. They generally get in the way of active outdoor work, however, and—by catching and tangling obstacles—

they can be a positive hazard in some activities, like skiing and snowmobiling.

FOOTWEAR

Cold feet are the winter outdoorsman's single toughest problem. Your feet, which are subject to cold in the best of circumstances from being at the body's extremities, are especially hard to keep warm outdoors. There, they must be in contact with the cold ground and they are apt to get wet, from perspiration or from outside damp, or both. Once they get wet, feet can't dry as easily as other body parts, for tight, water-resistant winter boots seldom permit free evaporation.

The virtues of good winter footwear are water-tightness, warmth and light weight. Rubber boots with felt linings, leather boots with felt in-soles, rubber lower boots with leather uppers stitched on, all are intended to be as warm and dry as possible while staying light. Boots like these are usually available for $30–$40 a pair.

A less expensive solution to the problem of cold, wet feet, and one that works very well for many winter activities, is a nine- to ten-inch-high felt boot worn inside a heavy rubber or four-buckle "arctic" overshoe. When this combination is supplemented with interchangeable felt in-soles, it is very warm and can be kept dry simply by changing in-soles daily. The popular Sorel boot is similar to the felt boot–rubber overshoe combination, but more expensive. In Sorel boots and other felt-rubber combination boots, the felt liner should extend above the top of the leather or rubber top. This arrangement allows the liner to act as a wick, drawing moisture up out of the bottom of the boot.

Before you pull on your boots for a winter expedition, put on an extra pair of socks if the weather is at all cold. The layering principle is at work again here: two pairs of socks allow for an insulating air space between them. Make sure when you buy boots that they are roomy enough to accommodate your feet comfortably in two pairs of socks (it might be well to wear two pairs of socks when you are trying on boots in the stores). Lightweight socks—cotton or silk—should be the first pair to go on, then heavier wool or thermal socks. Natural materials are superior to synthetic for winter socks, as they are for other winter clothing items, because they allow for freer evaporation of moisture.

When you are well shod and ready to go outdoors, pause for a minute on the doorstep to let your house-warmed boots chill. This will prevent their picking up snow as you walk, and will postpone (if not prevent) damp feet.

HAND- AND HEADGEAR

It is not always possible to wrap your hands as warmly as they should be in winter, for they must be free to move and grasp. Nevertheless, to avoid frostbite, hands must be covered, preferably with mittens. Gloves, though convenient and unconfining, are nearly useless for warmth. Each finger of a gloved hand is encased in its little sleeve, where it quickly grows cold. In mittens, your fingers keep each other warm. The warmest arrangement is a light cotton or woolen pair of mittens worn under heavier wool, leather, fur or insulated fabric outer mittens. Cotton gardening gloves also work well under woolen mittens, though woolen liners under leather outer mittens are less slippery. If you must wear gloves, silk or cotton under gloves, tight fitting, under a larger pair of leather gloves give added warmth.

Gauntlets are helpful in protecting your wrists and in keeping snow out of your coat sleeves. You can add gauntlets to your own mittens or gloves by cutting the wrists off an old pair of woolen mittens and stitching them on inside your regular over-gloves.

To prevent loss of mittens or gloves, an old mother's remedy is effective. Fasten a long string to one mitten, run its free end through both sleeves of your coat, then fasten it to the other mitten. This will insure your mittens never get away—an important matter on a long winter outing.

On extended outings, take along an extra pair of mittens. You will be glad to change into them when the pair you started with gets wet. Battery-operated hand-warmers may also be worthwhile.

As we have seen, body heat can be easily lost at the head. In any but the mildest winter weather it pays to cover up. A *tuque* or knit stocking cap, wool, works well in most winter weather. It is warm and allows moisture to evaporate, and you can pull it down over your head and ears to give extra covering. For colder weather, and especially for protection against wind, a balaclava

Winter headgear. Clockwise from left: balaclava helmet; ski mask; stocking cap.

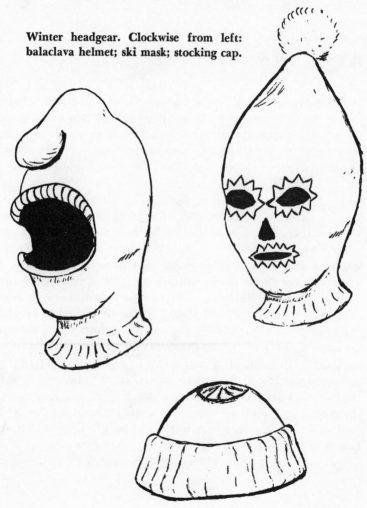

helmet or ski mask is a good buy. Also knit, these can be pulled right down over the face (holes or windows are let into the knit for eyes, nose and mouth). Visored caps are helpful in outdoors work on days when the sun is bright.

CLOTHES FOR SNOW SPORTS

Certain winter recreational activities have their own clothing requirements—not all of them related to warmth alone. The layering principle holds good for warm dress in all these activities, but outer wear may differ among them.

Cross-country skiing is so strenuous that there is little need for warm dress until you stop and rest. For the vigorous skier and racer, cotton underwear, knickers or other pants, and a light wool shirt—with gloves and an ear-protecting headband—should be enough clothing.

More leisurely cross-country skiers need warmer clothes for their less active touring. Again, wool can be recommended, especially as, unlike cotton, it keeps much of its insulation value when wet. Woolen whipcord or gabardine are best for ski touring pants, as they allow snow to be brushed off easily. Shirts and pants for ski touring should be loosely cut and unconfining. Lightly insulated, hooded jackets with double zippers that work up or down are convenient. Large jacket pockets are convenient for carrying food, extra gloves, or other equipment. A warm vest or sweater under the jacket may be needed in colder weather.

A small back-, belt- or fanny pack can be an important piece of ski touring wear for long outings. Use the pack for provisions, sweaters, gloves and hats. When buying a pack for skiing, try it on and be sure it doesn't interfere with your arm movement.

Streamlined, wool-and-synthetic fiber one-piece suits are now available for cross-country skiing, and are gaining popularity, with the sport itself. Worn with insulated underwear, one-piece suits can be warm (though not as warm as wool); but their chief advantage is that they offer less wind resistance than more conventional clothing, and therefore help in running fast.

The best clothing for the **downhill skier** can depend on several factors, including temperature, wind conditions, and the proficiency of the individual skier. Long underwear is always necessary. Racers also wear streamlined synthetic suits, lightly filled with fiber. Padded gloves and a snug hat, with fog-proof goggles and a crash helmet, complete the downhill racer's outfit. Racing ski boots are higher and tighter than other downhill ski boots, and far more expensive.

Less ambitious downhill skiers usually wear warm, bulky sweaters, and down- or fiber-filled parkas on really cold days. An insulated vest is a convenient article of clothing for skiing, as is a warm-up suit for the coldest weather.

Clothing for **snowshoeing** can be similar in most respects to that for ski touring. Felt boots with two pairs of socks are especially comfortable with snowshoes, as are the traditional Indian moccasins specifically developed for snowshoeing. Snowpac or Sorel brand boots are also good for snowshoeing, though they are more confining than felt boots.

For winter **joggers,** a sturdy set of woolen underwear, and a wool or cotton turtleneck topped by a wool- or fleece-lined hooded sweatshirt and a light windbreaker can be recommended. Wool or nylon warm-up pants are also comfortable. A ski mask or balaclava helmet are often needed for the jogger's face protection.

Snowmobiling requires very warm dress. Although driving a snowmobile is a more strenuous activity than it might appear to non-riders, the high speeds make for wind-chill conditions that demand extremely warm clothing. Under-layers of clothes for the snowmobiler can resemble those of other active outdoorsmen, but the outer covering must be more complete, to resist wind and hold body heat.

Cold-weather bib overalls, made of smooth-finish synthetic fabric and insulated with down or polyester fiber, are preferred by many snowmobilers. They often zip down the sides of the legs as well as down the center, providing for ventilation, and for easy access and removal. In warmer conditions, the zippers can be opened to allow cooling. These overalls are usually worn with a short wind-proof jacket, and they have the drawback that snow kicked up by the snowmobile can find its way between jacket and overalls.

A better choice in snowmobiling clothing is the newly-popular, one-piece jumpsuit. Resembling mechanics' coveralls, jumpsuits are usually made of smooth-finish synthetic fabric insulated with down or polyester fibers. Although much too cumbersome for most winter activities, jumpsuits seem ideal for snowmobiling (they are expensive, costing around $50). Refinements on the basic snowmobiler's jumpsuit may include padded or reinforced knees and seats, zippers at the legs to allow easy access and ventilation, straps under the feet to keep the legs of the suit from working up out of the boots, and reflective patches on back and arms for safe night riding.

Jumpsuits should fit loosely at shoulders, knees and elbows, to allow for the freedom of movement essential to applying "body English" in steering a snowmobile. Like other winter outer wear, however, the suits should fit snugly at neck, wrists and ankles.

Footwear for snowmobiling should be the warmest possible. Felt boots in rubber overshoes and Sorel boots are a good choice, but many snowmobilers favor specially-designed snowmobile boots. These employ a felt inner layer surrounded by an outer shell of rubber and nylon. The rubber sole has a deep tread to give good footing on icy pedals and running boards. Above, a nylon sleeve or high top zips up the ankle and closes with a drawstring to keep snow from sifting down into the boot.

The best headgear for snowmobiling is the typical motorcyclist's helmet, combined with a pair of impact-resistant snow goggles. On really cold days, pull on a wool knit helmet or stocking cap under the crash helmet. The stocking cap pulls down over the back of the neck and around the ears. Ski masks are also good snowmobiling gear for bitter days. Snowmobile drivers should avoid hoods, which can cut side vision.

For hand protection, snowmobilers should wear double mittens, the outer pair of which has an ample gauntlet which can be drawn up over the sleeves to keep out blowing snow. In some snowmobile outfits, gauntlets attach to jumpsuit sleeves with Velcro strips. Although mittens, being warmer than gloves, are better than gloves for snowmobile riding, a pair of gloves should be in the snowmobiler's pocket for use when the machine needs tinkering.

For snowmobilers who ride on frozen lakes, ponds and streams, it is a wise precaution to add to the standard gear a life jacket and a pair of short screwdrivers. The screwdrivers should be carried in sheaths sewn onto the life jacket. If the machine goes through the ice, the rider can use the screwdrivers as picks to punch into the ice in pulling himself to safety.

CLOTHES FOR OTHER WINTER ACTIVITIES

Among the popular winter occupations which require special thought to be given to clothing are hunting, ice fishing and attending at outdoor sports events.

Hunters are abroad in all sorts of winter conditions, and their activity levels are equally various: they may sit immobile for hours, or they may have ground to cover. For the least active hunters—waterfowl gunners, for instance—warm insulated parkas may be indicated. For most others, however, the key is flexibility; dress in light layers and add or remove clothing as needed.

Hand covering is especially important for hunters, who must be able to work a gun or bow quickly and efficiently. Shooting mittens, which have a separate forefinger and the other fingers bunched, are a good buy for cold weather hunters.

Among the more sedentary winter activities, **ice fishing** and **attending outdoor events** have similar clothing requirements. Large, well insulated garments—parkas or snowmobile jumpsuits —are well fitted for each. In both, too, footwear should be warm and dry but need not be especially light. The heavily insulated rubber boots developed for the armed forces (called "Mickey Mouse" boots) are often used by ice fishermen and spectators, though they have the disadvantage that they do not permit free evaporation of moisture and therefore can become damp from perspiration. Felt boots worn with rubber overshoes are probably a better choice in footwear for sedentary winter occupations.

For spectators and winter fishermen (and others), body insulation can be improvised through the use of many materials, including newspaper and plastic bags. Standing on a section of newspaper spread on cold ground or concrete can help keep your feet warm, and pieces of newspaper stuffed into the waistband or buttoned under a jacket can provide an extra layer of warmth. Plastic, too, makes good body insulation. Used to wrap cold hands and feet, plastic freezer bags can compensate for forgotten mittens or extra socks.

Battery-powered hand warmers are ideal for less active winter outdoorsmen. As for liquid warmers, coffee, tea and cocoa are fine, but liquor is to be avoided if you will be outdoors for more than a few minutes after you take a drink. Alcohol's immediate effect is to warm you, but its ultimate effect is to draw warm blood from the center of your body to your head and extremities where it can cool quickly, robbing you of body heat.

CARE OF WINTER CLOTHING

The variety of natural and synthetic materials used, often in combination, in winter clothes can make it hard to know how best to clean them. Most manufacturers print cleaning instructions on garment labels, but labels can get lost, and second-hand clothing may lack labels. When in doubt about how to clean a garment, wash it gently in cool water, by hand or in a washing machine (wash all good woolens by hand). **Down-filled garments** often require dry-cleaning; when they can be washed, they need a mild detergent.

Don't put **woolens** in your clothes drier: lay them on a towel over a radiator or near a hot air register so they will dry slowly. For down-filled garments, put them in a drier at moderate heat, and put a sneaker in the drier with the clothes (it will toss around with them and help fluff up the down). If you have no drier, shake the garment up by hand from time to time as it dries.

In homes without clothes driers, clothes can be dried on or near radiators and registers, on racks near fireplaces and stoves, or even in a sunny south window. Remember, however, to avoid excessive heat in drying down and woolen items.

Don't forget your clothesline in winter. Hung on the line on a bright, sub-zero day, sheets and towels will freeze stiff, but they will dry anyway—and the freezing and open air drying seems to make some articles fluffier than any commercial fabric softener could. Don't hang woolens or linen articles on the line to dry, however.

Repair of most winter clothing is usually a simple matter of needle and thread. In the case of some synthetic fabrics, very satisfactory temporary repairs can be made with tape. Stuff protruding feathers and quills back into down-filled garments instead of pulling them out. When the time comes to store winter garments for the summer, make sure to provide moth balls for woolens, and try to hang down-filled garments up without crowding them against other clothes. This avoids compressing the down; compression reduces its insulation value.

Leather boots and shoes need more attention in winter than in other seasons, for they are subject in winter to repeated wetting and drying and exposed to road salt and sand. Treatment with silicone preparations, neats foot oil, and mink or castor oil can help waterproof leather and keep it supple, prolonging its life (silicone products are recommended for suede and other rough-textured leathers). Thorough cleaning with saddle soap will also help preserve leather. In winter, leather boots and shoes should ideally be cleaned and treated with waterproofing and softening preparations at least once a week.

Like wool and down, wet leather should never be dried in intense heat. Let your wet boots dry slowly over a register or at a distance from a stove or fireplace. Newspapers stuffed loosely in wet footwear help remove inside moisture and hasten drying, while holding the shape of the shoe or boot.

GOOD READING ON WINTER DRESS

A good source of published information on what is available at what price in the way of winter clothing is retailers' catalogs (read with reasonable skepticism where appropriate). *Two especially informative catalogs are the monumental volume issued by* Sears, Roebuck and Co., *and the excellent catalog of* Eastern Mountain Sports, Inc./ (EMS). *Both include brief texts on the materials and construction of their products. The Sears catalog is available from local Sears, Roebuck catalog shopping outlets free to Sears customers. The EMS catalog is available from Eastern Mountain Sports, Inc., 1041 Commonwealth Ave., Boston, Massachusetts 02215 (cost, $1.00).*

Sew-it-yourself kits for making down-filled garments of all kinds are available from Frostline Kits, *452 Burbank, Broomfield, Colorado 80020; and from* Holubar Mountaineering, Inc., *P.O. Box 7, Boulder, Colorado 80306.*

7 Winter Health and Diet

IS OUR HEALTH BETTER, or worse, in winter? The best answer may be that the season *need* not make much difference, though it often does. Both folklore and medical science agree that cold *per se* does not produce illness. On the contrary, an old north-country saying holds that "a green winter fills the grave-yards," suggesting that a full winter season with a good snow cover is more healthful than an "open" winter with little snow and seesawing temperatures.

Nevertheless, though winter weather may not be in itself a threat to health, there are health problems which may be more troublesome in winter than they are in milder seasons—not because they are caused by cold, but because cold weather makes them more severe and makes us more susceptible to them. Therefore, while your regime for maintaining health may not need major changes to meet winter conditions, the colder months will require some additional precautions if you are to stay healthy, especially if you are subject to certain medical problems in the first place.

Cardiovascular troubles are aggravated by exposure to cold weather because chill puts extra demands on the circulatory system to maintain body temperature. Persons with heart disease should avoid any sort of heavy exertion at low temperatures. Snow shoveling, to take a prime example, can put a dangerous strain on a weak heart (and can cause lower back injuries even in healthy people, see Chapter 4). No one with heart trouble, high blood pressure, or other circulatory ailments should expose himself to rapidly alternating extremes of temperature (for instance, in a sauna bath).

COLDS AND INFLUENZA

Chronic respiratory problems, and colds and influenza, are also likely to be worsened by exposure to very cold air, as are all inflammations of the moist membranes of the respiratory system. Colds and influenza are highly contagious diseases, and 'flu, especially, is associated with cold weather. Although it can be prevented to some extent by inoculation (as colds cannot), 'flu is a dangerous illness which can weaken the system, allowing other, more serious, ailments like pneumonia to get established. 'Flu is a particular winter threat to elderly persons and persons with chronic respiratory problems, each of whom should, at the onset of winter, get medical advice on inoculations and other 'flu preventives.

Your physician can tell you if particular 'flu shots are recommended, and when they will be available; or you can ask your local public health officials. In large cities such officials are reachable at the Public Health Service, listed in the telephone directory under U.S. Government, Department of Health, Education and Welfare. In some states with relatively small populations the listing is under the Health Department of the (state) Human Services Agency. In rural areas look under the specific township or county, for the Health Officer and the Public Health Nursing Service.

These agencies can also tell you of scheduled clinics for immunizations or preventive check-ups for all ages and for a variety of disabilities.

'Flu and colds may have similar early signs. If you feel a cold coming on, don't go out of doors in very cold weather to try to "work it off." A physician at one of the largest ski resorts in the Northeast reports that the most common ailment he treats is sore throats, suffered by holiday-makers who hope that a day on the slopes will clear up a mild congestion.

FROSTBITE AND HYPOTHERMIA

While colds, influenza and similar ailments can plague snow-country winter residents of whatever habits of life, frostbite and hypothermia are most apt to trouble outdoorsmen. Nevertheless, while these afflictions are produced by cold and snow, neither requires extreme conditions for its occurrence: you can get severely frostbitten at a football game as well as you can on a wilderness expedition.

Frostbite is actual freezing of skin and other tissue, blood, even bone. One of the most dangerous aspects of frostbite is that few people know how to treat it. For many years the standard treatment was to rub frozen parts or areas with snow. It was believed—wrongly—that this chafing action with a cold substance was the best way to thaw affected parts. Much permanent damage resulted from this treatment. Today, most authorities believe that rapid warming, without chafing, is the best treatment for frostbite.

Superficial frostbite, or "frostnip," can frequently happen to face skin exposed to low temperatures and high winds. Frostnip can best be detected by sight: white patches appear on frostnipped skin. When a group of people is exposed to potential frostbite conditions, every member should make a special effort to watch others' faces for signs of frostnip. Beards are a hindrance here, as they conceal but don't prevent frostnip (they can also trap condensation, which can freeze to produce an ice mask, thus promoting frostbite). Another method of detecting frostnip is to make faces by distorting your features. This makes it easier to notice, or feel, numbed areas of skin.

Frostnip is really no more serious than a mild sunburn, but it is important that it be treated immediately to prevent its developing into severe frostbite. The best way to thaw a face frostnip is to hold a warm hand over the affected area. When the white patch of frozen skin turns red, it has thawed. Frostnip in the fingers, which often appears first as white spots at fingertips, can be thawed by making a fist inside your mittens or by holding your hands in your armpits. Frostnip in the toes is more difficult to discover and treat. If there is any lack of sensation in your toes, remove your boots and hold frostnipped toes in warm hands or place them against another person's midriff.

Following immediate first-aid treatment of frostnip, a thorough warming-up session is in order. If you cannot get indoors, start a fire and make hot drinks—even a drink of hot water will help. Remove and dry damp, cold clothing.

Deep frostbite is a severe injury. The affected part, in addition to being white, is hard to the touch and has none of the resiliency of normal flesh. The best first aid for deep frostbite is to make all haste to some medical facility or to good shelter. Traveling to a place where help is to be had is less dangerous for persons suffering from deep frostbite than are inexpert attempts to thaw frozen parts under poor conditions.

Current medical practice indicates that the best way to thaw deep frostbite is by rapidly re-warming frozen parts in water heated to a temperature of 90 to 106 degrees Fahrenheit. After thawing, frostbitten areas should be treated in every respect like wounds, to avoid infection and further damage. If you must re-warm a frostbite victim without medical supervision, get him to a hospital as soon as you can after the frostbitten flesh has been thawed. Tissue damage from deep frostbite is often extensive. Even in relatively minor cases, injured parts may be sensitive to cold for years.

Hypothermia is a more common and more serious threat to winter outdoorsmen than frostbite. Although it is more prevalent in winter than in other seasons, hypothermia can occur at any time of the year when a person becomes so chilled that his internal body temperature drops significantly below normal.

Hypothermia's first important symptom is violent, uncontrollable shivering. The victim generally is clear headed at this stage, and can aid in his own recovery. Get him into a sleeping bag which has been warmed by someone else, or take other steps to warm the victim from without. When the shivering has stopped, warm drinks will usually bring the hypothermia victim back to normal.

In more advanced cases of hypothermia, the victim may seem drowsy and listless, as though he had been drugged. He cannot think clearly, and has evidently lost the will to survive. First aid in this case should be directed toward getting prompt medical attention, while keeping the victim in a neutral environment which is neither too warm nor too cold (wrapping a victim of advanced hypothermia warmly in blankets or a sleeping bag may prove fatal). Severe cases of hypothermia require careful warming of the body core under medical supervision in order to avoid pernicious changes in blood chemistry.

The current wide interest in winter camping, which may lead inexperienced people to take on outings in conditions they are not prepared for, has made hypothermia a real threat. The best way to prevent its threatening your winter camping is to have ample food, warm clothing, equipment and fuel. Above all, plan your trip carefully in all details, and use judgment in implementing the plans: if you find things much rougher than you expected, go home, and try it again another time.

OTHER WINTER COMPLAINTS

Not so serious as frostbite or hypothermia, but still troublesome, are a number of winter-related ailments including snow blindness, winter sunburn, chilblains and certain "occupational diseases" of snowshoers and snowmobilers.

Snow blindness may be a problem especially in late winter, when the mounting sun, its light reflected by the snow, can produce light strong enough to temporarily injure eyes. It is the ultraviolet component of sunlight that causes snow blindness. On bright, cloudless winter days, especially at higher elevations, unprotected eyes may begin to feel gritty. If exposure to strong sunlight continues, the eyelids will swell shut, causing temporary blindness.

The usual treatment for snow blindness is rest, with cold compresses placed over the eyes. There are generally no complications after recovery. Reddish- or brownish-tinted sun glasses, or improvised paper or cardboard snow goggles—opaque spectacles with slits cut into them—can prevent snow blindness. Short of true snow blindness, the glare of the unshaded winter sun off snow and ice can cause intense discomfort and impair vision. Dark glasses or snow goggles (which may be more effective than dark glasses in reducing glare) should be a part of any winter outdoorsman's equipment.

The same ultraviolet light rays which cause snow blindness can also cause a severe **sunburn;** especially on broad snowfields at high altitudes. Although the danger is greatest on sunny days, there is a real possibility of being sunburned on cloudy days as well. For protection against sun and wind burn there are many patent "glacier creams" (active ingredient: para-aminobenzoic acid or PABA). If none of them is on hand, petroleum jelly will protect exposed skin perfectly well, though messily. Be sure to apply skin creams liberally, coating the skin around your nostrils and under your chin, where sunlight reflected up from snow can burn you as badly as direct sunlight can.

Chilblains, another winter ailment, are inflammations of the skin of hands and feet repeatedly exposed to cold and wet. A chilblain remedy favored by farmers is the medicated salve used to soften and heal the udders of milch cows in very cold weather ("Bag Balm").

All winter activities and sports no doubt have their associated strains, sprains and other woes for certain participants. One common occupational complaint is *mal de raquette* (loosely translated: "snowshoer's foot"), which plagued the legendary *coureurs de bois* of yesteryear on their monumental snowshoe journeys. In *mal de raquette,* tendons in the ankle and foot become painfully inflamed. Rest and staying off snowshoes for a while should relieve the condition. Modern-day *coureurs de bois* who do most of their traveling by snowmobile avoid *mal de raquette,* but may incur a more serious handicap. Prolonged exposure to loud engine noise can cause premature deafness among snowmobilers. Take care that your snowmobile has a good muffler, or wear ear plugs on the trail.

EVERYDAY SAFEGUARDS FOR GOOD HEALTH

To avoid winter colds, 'flu and other illnesses a good diet, and a few commonsense safeguards are necessary.

Avoiding contagion entirely is next to impossible, since virtually none of us lives or works in isolation. Good sense, however, dictates here that persons with sore throats, fever, persistent cough, etc., should, if possible, stay home and take care of themselves and keep from spreading infection. People for whom such ailments would create undue strain should be wary of attending gatherings where contagion is likely to exist—and that means most any gathering.

Avoid chill. Take the time and thought to dress warmly. The quick dart outdoors when you don't bother to put on a warm jacket or waterproof footgear, failure to change out of damp clothing after an outing, spending time in a drafty or unheated area of the house without dressing warmly—all such lapses of vigilance can result in a chill which may reduce your system's resistance to infection.

Get plenty of sleep to preserve health. Fatigue increases the effect of a chill. A hot bath or shower, warm sleepwear, and a bed made cozy with a hot-water bottle or electric heating pad can work wonders for morale as well as for physical well-being. A small slab of soapstone or a brick, heated, and encased in a bag made from old turkish towels, produces an incomparable island of warmth in the foot of a cold bed. And a mug of hot lemonade, especially if honey is used to sweeten it and soothe a scratchy throat, makes a restorer that is hard to beat.

Make sure your house has **adequate ventilation,** especially if you use a fireplace or fuel-burning space heaters. Stuffy, overheated air can promote the spread of respiratory infections, and fires in unventilated rooms can lower the oxygen content of the air, while adding deadly carbon monoxide. Adequate ventilation is especially important in sleeping quarters, particularly in houses which use coal-burning space heaters. Burning coal produces more carbon monoxide than other fuels (see Chapter 3) : *never sleep in a room with a coal fire.*

Ventilation is also related to warmth. Old-time cooks who used wood-burning ranges would always air out the oven before they boosted the fire for baking. They contended (and from my experience they were right) that fresh air heats more quickly than stale air does. The same theory is applied by the housewife who airs a room thoroughly in order to get the most comfort from the heat in it.

Airing a room will also increase the relative humidity, though only temporarily. **Boosting humidity** indoors creates a more healthful environment. You can use any of the measures discussed in Chapter 3 to add humidity in the regular course of indoor activities. Special electric vaporizers are on the market for use at night in sickrooms to help people suffering from respiratory infections. Humid air relieves congestion and irritation in sufferers' lungs and sinuses. One type of vaporizer disperses cool water by blowing it out into the room as a fine mist; another boils water, thereby producing a mild jet of steam. A tent of sheets, rigged over the bed to contain the steam near the patient, will increase the effect of the vapor. Use cotton bedsheets, which absorb condensation. *Never use plastic sheeting of any sort:* it can fall over a sleeper or invalid and suffocate him.

Exercise is as important to physical health as it is to keeping happy. The next chapter describes a range of outdoor recreational activities. In addition to these, there are, ready and waiting, more impromptu brief activities which combine fresh air and physical movement to aid in good health. Get children out of the house in a snowman-making project (best done with rather wet snow), or show them how to make "angel wings" in dry, new snow by lying on the snow and moving both arms in 180-degree arcs, leaving swaths which resemble unfolded wings. Older people, or anyone else who fears a fall, can still enjoy winter walks by strapping "ice creepers"—simple mountaineers' crampons—under the soles of their footgear, or using a walking stick furnished with a nonslip tip for use on ice. Busy householders can make the most of routine outdoor chores like splitting and fetching wood, replenishing stores, hanging out the wash.

When outdoor exercise projects fail, there are always calisthenics. Set aside a particular time each day. Open the room to let in fresh air and lower the temperature, and do a series of exercises: bending, stretching, twisting, improving the tone of abdominal and lower-back muscles with sit-ups and mild push-ups, etc. Repeat each exercise only a few times in the early stages;

begin each stint with gentle, easy movements, then peak with more strenuous ones, and taper off with gentle exercises again. Avoid "ballistic" movements like bouncing down in deep knee-bends or flinging down to do toe-touches; and always keep knees slightly flexed during sit-ups and the like.

THE WINTER DIET

Good nutrition in winter is the same as in any other season: it consists in the right balance of protein, fats, carbohydrates, water, minerals and vitamins. Also as at other times of the year, volume of food intake and greater emphasis on particular foods depend on the intensity of one's physical activity and on the availability of various foodstuffs.

The **four main food groups** necessary for health are the meats (including poultry and eggs) and the protein-rich dried legumes like beans, peas and lentils, as well as nuts and nut butters; the vegetables and fruits; the dairy foods—milk, various cheeses, yogurt, ice cream, etc.; and breads and cereals. Today's nutritionists are de-emphasizing meat and eggs in the average American diet, suggesting instead that protein be taken more from vegetable and dairy sources. They are also recommending that we increase our intake of whole-grain cereals and enriched breads, and reduce our intake of sugar.

Mineral and vitamin dietary supplements generally are not needed during the winter if the daily diet is varied and well balanced. A possible exception to this rule is Vitamin C, which may be supplemented to help combat respiratory infections. Even here, however, an extra glass of fruit juice during the day effects the purpose of the Vitamin C supplement, and is more nourishing than a pill or capsule; juices have the added virtue of increasing the consumption of fluids, which is an important part of the regime for getting over many illnesses.

Certainly you should ask your doctor's advice before undertaking a significant program of mineral and vitamin supplements.

Professional advice is especially important where particular vitamins are concerned, for nutritional research is constantly discovering unwanted side effects from excessive intake of otherwise beneficial vitamins. For example, an adequate amount of Vitamin D, the "sunshine" vitamin, occurs in fortified milk; too much Vitamin D can overload blood and tissues with calcium. Even Vitamin C can be overdone: too much of it causes loose bowels, especially in older people.

Any list of nutritional values of food from the U.S. Department of Agriculture, or the Department of Health, Education and Welfare or the Food and Drug Administration, will help to insure a balanced diet, and many such pamphlets are available free from co-operative extension service County Agents.

Gaining weight in winter, for most people, is a result of increased intake of high-calorie foods without increased exercise to compensate. In general, the snowtime diet is likely to be relatively high in carbohydrates—which are the prime energy foods—and in fats, which also supply energy, and help to fight colds. Carbohydrates and fats are both high in calories. It is failure to burn up extra calories that causes many people to gain weight in winter. The logger-woodcutter and the cross-country skier or snowshoer can stay trim despite second or third helpings of macaroni-and-cheese and apple pie; the seasonally inactive person cannot. (Shivering burns calories, too, but there are pleasanter ways to go about it.)

Another factor in winter weight gain is the modern system of eating the day's largest meal in the evening, when there is little opportunity to work off some of its effects before going to bed. There is a great deal to be said for the old-time program of having a solid breakfast, a hearty midday dinner, and a lighter supper at night.

The importance of breakfast is a basic consideration at any time of the year, but especially so in cold weather. A hot, nourishing meal in the morning, eaten in cheerful company, provides energy and the morale to sail through activities at school, in the office, around the house or out of doors. People who habitually can't eat heartily first thing on arising can follow the example of the farmer who completes a round of chores before sitting down to a good breakfast.

Breakfast need not be the storied type of meal which includes ham and eggs, hash-brown potatoes, pancakes and a piece of pie. Many persons either "can't face" eggs in the morning or are limiting their consumption of eggs to avoid cholesterol; for them a well-cooked whole-grain or enriched cereal, fruit, perhaps a couple of slices of crisp bacon and certainly a glass of milk along with their coffee or tea will offer excellent nourishment. Hot cocoa is a good beverage for youngsters on their way to school or to play in the snow. Adults can fill a nutritional gap with a glass of milk, which is high in protein and is one of nature's best energy-boosters.

Some pleasant breakfast variations in the morning are cornmeal mush in place of hot wheat or oat cereal, slices of fried scrapple (cold wheat or corn mush that contains scraps of meat, well seasoned) or fried cornmeal mush in place of pancakes or french toast. Cook up cornmeal as for cereal, then put it in a bread pan to chill and set; slice it, dredge each slice in flour, and fry the pieces in hot butter, margarine or bacon fat, and serve with syrup or honey. Another food item that is a great favorite is fried bread. It's simple to make: just slice a home-baked loaf—rye or whole-wheat bread is best for this—quite thick; over medium heat melt butter just to smoking in a heavy skillet, put in the bread and reduce the heat. When each slice is nicely browned on one side, flip it over to brown on the other, and eat it hot from the pan with maple syrup.

COLD-WEATHER DISHES

Particular cold-weather dishes are a matter of individual preference, of course, but I would like to note here a few novelties, revivals and personal favorites. The selection of dishes that follows is meant as a source of ideas only; for detailed recipes, see any standard cookbook, like *Joy of Cooking, Fannie Farmer* or *Better Homes and Gardens*.

One winter dish that I remember with great vividness is *créton,* as prepared by a French–Canadian bull cook called The Old Man Pé. *Créton* was made from pork scraps, cleverly seasoned with spices; commercial deviled ham is something like it. *Créton* is a great specialty of the home cooking of Québec, La Belle Province, but I have never been able to find written directions for making it.

One of the best uses for deviled ham, or *créton,* is in sandwiches. The Old Man Pé almost invariably packed several huge *créton* sandwiches into his dinner pail when he went out into the snowy woods to get up the following winter's fuel supply. He built a small fire near where he was chopping wood, and toasted those sandwiches for his noon meal.

Mention of *créton* is a natural introduction to an important part of the old-time winter diet—**fat.** Active people outdoors in cold weather usually crave fat. A staple part of the winter diet of the woodland Indians in eastern North America was a mixture of animal fat and maple sugar. As a youngster I used to enjoy eating lard by the spoonful. Now, however, I learn that with those spoonfuls of lard I was sealing my doom. According to medical research, our tendency to drift from an active outdoor life, like that of The Old Man Pé or the Indians, to a sedentary indoor life—combined with today's psychological stresses and other factors—has made animal fats in our present diet a leading cause of heart disease. This revelation has contributed to the demise of a great many of the older recipes cherished in the winter kitchen all over North America, which used lots of fat in one way or another.

One such forgotten dish is the substantial dinner of **fried salt pork** with boiled potatoes and milk gravy. The pork should have streaks of lean in the fat, and the pork should be parboiled to remove some of the salt before it is fried. The modern palate has lost much of its taste for dishes like this one, in part because several hours of heavy work in the cold air may be required to whet the appetite for them. **Pork and beans** with brown bread is another dish that has slipped well down in the popularity scale, doubtless for the same reason.

A more exotic fatty food, now seldom encountered, is **pem-**

mican, which was invented by the Indians of North America to be used on journeys of peril and hardship. The great virtue of pemmican lies in the fact that it is indestructible, even at extremes of heat and cold, and will remain edible for decades if it is put up correctly. In the great days of the fur trade, pemmican was stockpiled against starvation should normal food supplies run short, as they frequently did in the North.

Properly made, pemmican is about eighty percent fat and twenty percent lean meat. In former times the bison, moose or caribou furnished the raw materials for pemmican, but beef will do just as well. The leanest cuts of meat are dried at low heat until six pounds becomes one pound, then the dry meat is thoroughly pounded and shredded, and packed rather loosely into a waterproof and heatproof container.

The best fat to use for making pemmican is the hard beef or mutton tallow from around the kidneys (soft fats like pork lard or chicken fat are much too greasy for the purpose, and they will not stabilize the pemmican thoroughly). Chunks of this fat about the size of an acorn are put into a shallow pan and heated slowly until the fat melts away from the surrounding tissue. After bubbles stop rising, the melted fat is strained to remove tissue particles, whereupon the strained fat is re-heated to just below the smoking point and then poured over the shredded meat in the container to make an airtight seal.

Pemmican certainly is not a food to figure in the regular diet of contemporary people, but it does have specific applications as a light, convenient and nourishing food at the winter camp. I adapted it into my diet during the winter of 1971–72 during a prolonged bivouac, and found it to be just about right. On a daily ration of 1½ pounds of pemmican, an adult can maintain health and energy for periods of up to a month, perhaps longer. People doing hard physical work under extreme weather conditions develop a real affection for pemmican, though it will never become a favorite hors d'oeuvre at fancy cocktail parties.

In recent years **potatoes** have been accused of causing America's bulging waistlines, but they are only a scapegoat: the real culprits are rich sauces, cakes and sweets—and inactivity.

While potatoes are often improved by being cooked in the company of other foods, they are also very good when eaten "as is" after being baked or boiled—with the skins on, of course. Try a couple of small baked potatoes as a trail lunch. When they are taken warm from the oven, potatoes make good hand-warmers at first, and later on they are a neat snack, since their skin container can be eaten along with the contents.

Potatoes were a staple of the old-time North Country cookery. One favorite potato recipe is for the always popular hash-browns. Most cookbooks have directions for these, but for some reason their recipes never come out for me as well as the recipe used by my mother-in-law. In her hash-browns she always uses cold baked potatoes for the main ingredient, and she slices her onions very thin instead of chopping them. She melts bacon fat to cover the bottom of the skillet generously, and fries the sliced onions until they are transparent. Meanwhile she pares the cold potatoes and chops them small in a wooden bowl. When the onions are done she pours them and the grease over the potatoes in the chopping bowl, adds salt and pepper to taste, and then returns the combination to the skillet in a layer that should be at least one-half to three-quarters of an inch deep. This layer is as important a part of her method as using baked potatoes and sliced onions. She lets the hash-browns cook at least an hour—longer is better. When they are brown on the underside, she turns them to cook a bit longer, but not long enough to allow them to dry out.

Cabbage, cooked or eaten fresh from the autumn garden or from the root cellar, is very tasty and nourishing. One of its most delectable transformations is sauerkraut. As sauerkraut, cabbage will keep for a couple of months right in the crock where it was made, and, if canned properly, it will keep indefinitely, to provide the foundation for many an elegant winter meal. Methods for making and canning sauerkraut are given in *Putting Food By,* by Hertzberg, Vaughan and Greene.

One favorite sauerkraut dish is made by cooking it up with pork chops and serving with baked potatoes. The important part of the whole process is to cook *slowly,* so essential juices can mingle together. Sear the pork chops in an uncovered dutch oven until the fat has been tried out, then slide the heavy pot over to a cooler part of the stove, or turn the heat down. Put freshened sauerkraut on top of the pork chops and replace the lid. After an hour or so, take the lid off and reverse the positions of the sauerkraut and chops—sauerkraut at bottom. Put the lid back on and let the combination cook over low heat for another hour (take longer if you can—you won't regret it). Potatoes can be baking in the oven meanwhile. The end result is delicious and exceedingly nourishing; the slightly acid flavor of the vitamin-rich sauerkraut counterbalances the pork fat and enhances the subtle aroma of baked potatoes.

In the old-time winter kitchen, the preparation of dishes was to some extent determined by what vegetables were in the family root cellar at what times (see below). A standby in early winter—when stored vegetables are at their prime—was the traditional Boiled Dinner of corned beef and vegetables along with Red Flannel Hash (cooked beets added to meat leftovers). Later on in the winter came pea soup cooked slowly along with a ham bone that hadn't been too closely trimmed of its meat. A good, thick, hot pea soup with lots of ham scraps was, and is, unbeatable for a noon meal on a cold winter day; or it could serve just as well at suppertime, depending on your family's eating habits. The best directions I know of for pea soup are given in Mrs. Rombauer's *Joy of Cooking*. Her recipe, by the way, is a good example of how to use a few spices and herbs to enhance a dish that many people remember as being bland and uninteresting.

Beef pot roast with Apple Crisp for dessert is another excellent mid-winter dinner. In this pot roast, as in many of the other dishes I have mentioned, long, slow cooking with selected vegetables adds flavor to compensate for using relatively inexpensive cuts of meat. *The Joy of Cooking* has good directions for preparing pot roast, but in our kitchen we vary them a bit by cooking the potatoes alongside the meat with the other vegetables. Toward the end of the cooking period, we spoon a little of the fat over the potatoes from time to time to help them brown.

My wife's recipe for apple crisp makes a perfect dessert to follow pot roast. To make her apple crisp, pare and slice six to eight apples and put them in a greased casserole. Mix together ¼ cup light brown sugar, ¼ cup white sugar, ¼ teaspoon nutmeg and about one teaspoon of lemon juice if the apples are sweet. Work this mixture in amongst the apples in the casserole. Then prepare a topping by combining ¼ cup light brown sugar, ¼ cup butter, one teaspoon cream and one cup flour. Spread this topping over the apples. Cover and bake in a 300-degree Fahrenheit oven for an hour and a half. Serve hot from the oven with thick, fresh cream.

Many north-country kitchens use **snow in winter cooking**. Unlikely though it may seem, snow can be an important ingredient in, among other things, griddlecakes and breads. Snow acts as a leavening agent, giving a remarkably good, open texture to baked goods.

One good snow-cookery item is cornbread. The ingredients are two cups of yellow corn meal, ½ teaspoon salt, one pinch of fresh black pepper, and four cups of clean, fluffy snow. The method: thoroughly mix the corn meal, salt and pepper in a large (ten-inch diameter) bowl. Grease an eight-by-eight-by-two-inch cake pan and lightly dust the bottom with flour. Then place the bowl with corn meal mix, cake pan, and a wooden spoon, a fork and a measuring cup outside or in a cold room to chill thoroughly. Heat the oven to 400 degrees Fahrenheit. Go outside and scoop up the four cups of snow and mix them thoroughly with the other ingredients in the bowl, using the wooden spoon and fork as if you were tossing a salad. Put the mixture in the cake pan without packing it down tightly. Immediately place the pan in the hot oven and bake about fifteen minutes, or until the crust is golden brown. Eat it with butter and maple syrup. Those who want to venture further into snow cookery might try the variation recommended by one author on camp cookery who suggests making pancakes of extra-thick batter into which a spoonful of snow is mixed for each cake just before it is put on the griddle.

Snow has other edible applications. "Vermont" (or "Minnesota," "Alaska," "Yukon," as the case may be) sherbet is simply fluffy new snow scooped up and eaten right out on the trail. For flavor, put two cups of new snow into a chilled bowl. Add ¼ teaspoon of vanilla, two tablespoons sugar and ⅓ cup whole milk or light cream. Stir these together in the bowl until all are well mixed, and eat immediately.

Another "sherbet" is particularly appealing to children in bed with colds. Pack freshly fallen snow firmly in a cereal bowl, and onto it pour ½ cup fresh orange juice or a smaller amount of partly diluted frozen juice concentrate. The youngster can start eating the mixture with a spoon, and end by taking the cold liquid through a straw.

STORING FOOD FOR WINTER

In snow country, the older farmhouses generally were built over large cellars with the thought that a good part of the below-ground space would be devoted to the economical practice of **root-cellaring.** Modern cellars are often heated and therefore unsuitable for root-cellaring. If your house has no ready-made root cellar space, however, you can adapt a cold garage or bulkhead for root-cellaring, or you can build an outdoor root cellar using a "cold pit" (see drawing), or a buried box or barrel.

The book *Putting Food By,* mentioned earlier, has instructions for creating a number of root cellars, with directions on humidity, temperature, and the handling of specific root cellar commodities.

In a root cellar, fruits and vegetables are held in a cool (32–40 degrees Fahrenheit), damp space to prevent freezing and retard natural decomposition. Along with boxes of apples and potatoes, prime specimens of beets, carrots, turnips and other garden products are stored in the root cellar, where they will remain in fine condition for months, providing an unbroken supply of nourishing food throughout winter and into spring and early summer.

While root-cellaring may be the most convenient and economical way to store some crops, it is also worth remembering

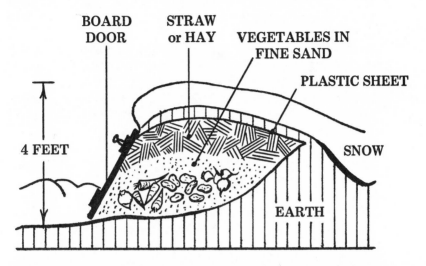

BOARD DOOR   **STRAW or HAY**   **VEGETABLES IN FINE SAND**

PLASTIC SHEET

4 FEET   **SNOW**

EARTH

"Cold pit" for outdoor vegetable storage in cold climates. A pit arrangement like this one gives better protection at very low temperatures (i.e., −20 Fahrenheit) than storage in earth mounds, cones, etc., above ground level. (From *Putting Food By*, Hertzberg, Vaughan and Greene.)

that some of the root vegetables can be "stored" right in the soil where they grew, in the garden under a protective layer of snow. Parsnips are definitely improved by overwintering in the ground, and Jerusalem artichokes make a nice crisp treat to have on the table the moment that snow is gone and frost is out of the ground.

Winter storage of food outside the root cellar calls for certain precautions, mostly concerned with avoiding freezing. **Canned goods,** including foods home-preserved in sealed glass jars, should be kept in a cool, dry, dark place where they cannot freeze. A preserving container's "headroom"—the small amount of space not taken up by food in the container—does not allow for swelling as the food freezes. If its contents freeze, the container's seal will break, even if the container itself does not, and the contents will spoil. **Dried foods** also should be stored in a cool, dry, dark place, and in containers which are insect- and rodent-proof. Large canning jars or commercial glass containers do well; so do large cans. It is important to have well-fitting metal lids for all such containers, because rodents seem to have no trouble gnawing through plastic lids.

Freezing food to store it should only be done in a good, modern food-freezer, in winter as well as in other seasons. Don't make the potentially fatal mistake of supposing you can freeze and store food safely just by putting it outdoors in winter. It is dangerously easy to overestimate the ability of bitter cold weather to act as a natural freezer for preserving food against spoilage. The fact is that there is virtually no habitable area of North America where the outdoor temperature is the unwavering zero degrees Fahrenheit (−17.8 Celsius) required for safe prolonged frozen storage of food.

Freezing stops the growth of bacteria that cause food to spoil, but the bacteria survive; when the temperature rises, they start to multiply again; with another sharp freeze they become dormant once more, but meanwhile the number of inactive bacteria has increased. A few more temperature fluctuations, followed by thawing at the equivalent of refrigerator or even room temperature, and the food carries a bacterial load which only very long cooking at quite high temperatures can render harmless.

In all freezing, follow these recommendations of food specialists in the United States and Canada: Do not store food above 10 degrees Fahrenheit (−12.2 Celsius) for more than a few weeks; Freeze food initially at −20 degrees Fahrenheit (−28.9 Celsius), then make sure it is held at a constant Zero Fahrenheit. Low temperatures this constant cannot be attained simply by exposing food to outside winter air, however cold. Even using all safeguards, moreover, fatty foods have only one-half to one-fourth the frozen storage life of non-fatty foods; and meats, poultry and seafood are especially likely to harbor food-poisoning bacteria, no matter how carefully they are frozen.

Do not re-freeze any food unless it has been thoroughly cooked since it was thawed. If, during cooking, frozen food foams unduly, or gives off an unpleasant odor, destroy it in such a manner that it cannot be eaten by people or animals.

SPROUTING

Sprouting seeds for "live" food during the winter is an age-old procedure which is becoming increasingly popular in North America, partly because it is easy, but mainly because sprouts are extremely high in nutrients, notably Vitamin C. Alfalfa, lentils, and mung beans are among the seeds most often used for sprouting. If you are a beginner at sprouting, start with mung beans—they are practically indestructible.

Sprouting must occur in the dark, and at normal room temperatures (60–80 degrees Fahrenheit). Seeds must be rinsed several times a day to avoid spoiling and off flavors. The simplest sprouter is a quart canning jar with several thicknesses of clean coarse-mesh cloth held over the mouth with a rubber band around the jar's neck. Soak seeds to be sprouted overnight, put them in the jar with the cloth in place, and rinse them with drinkable tap water. Lay the jar on its side in a dark cupboard or box, or cover it with aluminum foil pressed all around it. Inspect seeds morning and night, rinse them, and throw out seeds that have failed to sprout. Continue until your sprouts are the desired length and the sprouted seed has increased to double its original volume, or more.

In sprouting, there are three important precautions which must be kept in mind. *Use only* seed which is certified as free of pesticides or dyes, and therefore is declared to be edible (don't use garden seeds for sprouting); dried legumes sold for food in supermarkets—lentils, beans, peas, etc.—are food pure and can be sprouted. Don't overindulge in raw sprouts; one cup of most sprouted seeds contains twelve times the Vitamin C in a cup of orange juice, and a Vitamin C dose of such magnitude can cause loose bowels. Finally, *never* eat tomato or potato sprouts: both are poisonous.

GOOD READING ON WINTER DIET

United States Department of Agriculture
The following Home and Garden Bulletins of the U.S. Department of Agriculture (USDA) are available—usually free or at low cost—from the Office of Information, USDA, Washington, D.C. 20250. Refer to bulletin numbers.

No. 10   *Home Freezing of Fruits and Vegetables.*

No. 70   *Home Freezing of Poultry.*

No. 119   *Storing Vegetables and Fruits in Basements, Cellars, Outbuildings and Pits.*

No. 162   *Keeping Food Safe To Eat.*

Hertzberg, Ruth; Vaughan, Beatrice and Greene, Janet. *Putting Food By.* 1975. Brattleboro, Vermont, The Stephen Greene Press. Available in bookstores or from the publisher for $4.95 (paperback). Mass market paperback (Bantam Books) available for $2.50.

Munroe, Esther. *Sprouts To Grow and Eat.* 1974. Brattleboro, Vermont, The Stephen Greene Press. Available in bookstores or from the publisher for $4.50 (paperback).

Klippstein, Ruth N. and Humphrey, Katherine J. T. *Home Drying of Foods.* (No date). Information Bulletin No. 120, Division of Nutritional Sciences, Cornell University, Ithaca, New York, 14853. Available from the publisher for forty cents.

Watt, Bernice K. and Merrill, Annabel L. *Composition of Foods, Raw, Processed, Prepared.* 1963. Agriculture Handbook No. 8, USDA. Available from the U.S. Government Printing Office, Washington, D.C. 20402, for $2.00.

8 Keeping Happy

WINTER IS NOT, by tradition, a happy time of year: we may feel it as a season to be endured, not enjoyed. In the past, before the development of modern transport and communication facilities, winter's cold, snow and darkness brought a season of isolation, boredom and despondency. To some extent they still do, though good roads, the telephone, and radio and television enable us to avoid winter stagnation more easily than our forebears could. Nevertheless, the rigors of winter weather and winter's long nights can affect our spirits, and can produce that compound of boredom, torpor and depression we call "cabin fever" or "shut-in syndrome." Cabin fever, incidentally, is by no means confined to the American snowbelt. By other names, including "arctic hysteria," *'piblokto* (from Greenland), and *latah* (from Mongolia), winter's depression is known by all northern peoples. Generally, cabin fever is a temporary ailment, easily put to flight by a sunny winter day, a visit, or a long walk. Cabin fever can become a debilitating problem, though, especially if it is "treated" with alcohol or other drugs.

However your own spirits may be affected by the hardships of winter weather—however subject you are to cabin fever—a little thought will show you that our spirits should be prepared for winter as well as our bodies, our houses, our automobiles. In preparing our *selves* for winter, there is one principle to have in sight: *Don't hibernate.* Bears and groundhogs can sleep the winter away, but you can't—don't pretend otherwise. The most insidious effect of winter on our lives is the idleness it enforces upon us, an idleness or laziness that is mental as well as physical. Avoid

141

idleness of each kind: be as active as you can outdoors in winter, and keep your mind and attention engaged indoors.

In countries with some of the hardest winters on earth, people have apparently seen the wisdom of the principle that happy wintering is active wintering. In Lapland, for instance, the nomadic peoples traditionally gathered together in late winter with their reindeer herds for late winter festivals—festivals which included contests of skill and endurance and races of all sorts. Winter games have also figured in the traditions of Scandinavian peoples and North American Indians, among others. It is this determination to remain active in spite of winter's restrictions that we should imitate in meeting our own winters.

WINTERING INDOORS

Indoor activities which can help to keep us alert and immune from cabin fever are many and various. Take advantage of the fact that winter makes many outdoor jobs impossible. If your house needs insulation or any other indoor improvement, winter is the time to do the work. The recent spread of interest in crafts opens a vast field for winter activities. It is no coincidence that many of the "country" handicrafts now enjoying a revival, including wood carving, toy making, quilt making, dyeing, were originally winter work for pioneer farmers who had little spare time in other seasons, much in winter.

In spite of its shortcomings, which are as real as they are well known, it is appropriate here to say a good word for that widely maligned institution, Television. Many accounts of winter in small, isolated, and often poor rural communities in our grandfathers' times and before tell of an oppressive weight of boredom, idleness and depression. For these communities cabin fever was no joke. Television has probably done much to relieve the burden of winter on people in out-of-the-way country places far from more complex forms of winter entertainment.

Evening extension courses at the local high school or college, and correspondence courses, give opportunities for directed winter activity. Reading, writing, artwork and hobbies can all be recommended. Many people think they can draw, and many of them are right. Winter is a good time to find out one way or the other. Some collecting hobbies may have a seasonal element built in which favors indoor work in winter: collectors of the products of Nature's summer bounty of insects and plants, especially, can spend the winter ordering and studying specimens they have gathered in warmer seasons.

Since the weather is a subject of universal interest in the northern winter, winter is an appropriate time to launch yourself on a career as amateur weatherman. Sporting goods catalogs and dealers can provide the basic equipment, which includes barometer, thermometer, precipitation gauge and wind-speed meter; and there are several good books which introduce the beginner to the study of meteorology. For the beginner, the essence of the subject is keeping careful, systematic records. Meteorology can be a rewarding and useful hobby.

Games exist in an almost infinite number for our diversion on winter evenings. Board games, from the old favorites checkers and Monopoly to the newest micro-circuited, electronic tennis-video-simulator, can be sovereign for cabin fever. Parlor games and card games are even more various. At the end of the chapter are listed two books from this publisher which together include enough game ideas to furnish a century of winters.

WINTERING OUTDOORS: FROM SKIS TO SNOWMOBILES

Keeping active outdoors in winter is at the heart of enjoying that season and turning it to good account. Winter sports have grown increasingly popular in snow country, and the northern states and Canada are well provided with facilities for hockey, downhill skiing, skating, sledding and snowmobiling. The usefulness of these activities in overcoming the winter doldrums is obvious. In participating in any of them we meet winter halfway on its own ground, realizing that we may as well join winter—for we can't beat it.

In what follows, we will emphasize winter outdoor activities which do not require elaborate facilities or (in most cases) expensive equipment.

Cross-country skiing, in the United States, has grown in fifteen years from a relatively obscure offshoot of downhill skiing (in fact, historically, downhill is an offshoot of cross-country) to a sport of immense and increasing popularity. The new waxless ski bottoms have made cross-country skiing a more satisfying sport for beginners than it was when a knowledge of the different ski waxes demanded by different snow conditions was a prerequisite for an enjoyable day on skis. The bottoms of waxless cross-country skis have irregular surfaces, or mohair strips bonded to them, to grip the snow and allow the skier to stride. In deciding which type of ski bottom is best for you, rent or borrow different kinds of skis before you buy.

For most of us, cross-country's chief advantage over downhill is that it thrives away from the expensive, overcrowded "ski area," with its lift lines and three-dollar hamburgers. You can ski cross-country whenever there is snow: on bike paths, golf courses, country roads, in parks and campgrounds. If you prefer to ski on prepared trails, you can find excellent ones for cross-country in places where downhill skiing would be impossible, for the creation and maintenance of a cross-country trail is a simple matter compared to that of a down-hill slope.

Snowshoeing is as versatile a winter activity as cross-country skiing. Getting around on snowshoes is excellent winter recreation, which lends itself to many other outdoor pursuits. Some modern snowshoes use neoprene or rawhide webbing on an ash frame, and some use neoprene webbing on an aluminum frame; other models (slightly less expensive, in the $40 range) are all plastic.

Of the several styles or shapes of snowshoe, the best for beginners is probably the tail-less "modified bearpaw" design. Because it lacks a tail and is comparatively narrow, this snowshoe is often easiest to use in brush and on hilly terrain. Beginning snowshoers should find it helpful to use ski poles as aids while getting the feel of their snowshoes.

Snowshoe materials and styles: Left to right: new aluminum-frame model; standard bearpaw style; Maine or Michigan style; Green Mountain bearpaw.

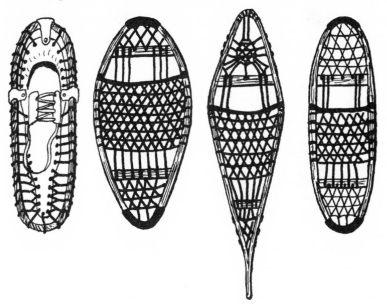

One especially appealing aspect of snowshoes is the possibility they offer for adapting a variety of outdoor games to winter. Try baseball, football or badminton on snowshoes: the workout, and the laughs, in all three will be increased when the fleet participants are wearing snowshoes.

Sledding is a "winter sport" which can be as simple or as elaborate as its followers want. You can slide downhill on a fancy bobsled, an old-standby "flexible flyer," or a plastic garbage bag. You can slide alone, with a friend, or with several friends on a toboggan. Don't select a sliding hill with a road at the bottom, and never ride on sleds being pulled along roads or streets by cars. Watch out, when you select a sliding hill, for snow-covered rocks, branches and barbed wire. Good downhill sledding is easier to come by than good downhill skiing, and can be just as exhilarating.

A variation on hill-sledding is afforded by the Finnish kicking sled (*potku kelka*) —a narrow, steel-runnered sled with a handlebar. The sled has a seat for a passenger, but the driver holds the handlebar at the back of the seat and stands on one of the runners. He uses one foot to push or brake the sled. Kicking sleds can be run along a level, but only on hard-packed snow.

Dog-sledding is another winter activity which is growing in popularity in the northern states and Canada. Raising sled dogs (huskies or samoyeds) and racing them attracts new enthusiasts each year. A good sled dog can pull 1½ times its weight; in racing, a dog team can maintain an average speed of fifteen miles per hour over a considerable distance.

Perhaps **snowmobiling** should not be included in a list of winter activities which don't require expensive gear, for snowmobiles are costly ($500-$2,000). Nevertheless, we include snowmobiling here because it is unlike any other winter activity, and seems to be enjoyed by many who don't go in for other winter outdoor activities. Snowmobilers are a loyal breed, and great organizers. There is hardly a village in snow country which doesn't have its snowmobile club. The clubs serve many purposes: they organize courses in snowmobile operation, safety and maintenance; and their members are often indispensable in winter rescue operations. Snowmobile clubbers, too, seem blessed with very charitable natures; the clubs sponsor many community betterment projects and have proven themselves excellent fundraisers for both public and private causes.

WINTER CAMPING

There is no better way to enjoy winter than by putting its hardships to good use in a winter camping expedition. In winter camping, the pleasures of camping out in milder seasons are enhanced by the business of finding shelter—and even comfort—in the snowy winterscape.

If you plan a winter camping trip, think ahead carefully, for the sake of pleasure, and safety. Read up on winter camping. Make sure you have adequate clothing for all degrees of cold and levels of activity (see Chapter 6). In the course of your trip your own activity level will vary from quite strenuous to (you hope) sleep, and the temperature may range between comparable extremes: dress accordingly, not forgetting extra clothing to change into.

A good **sleeping bag** is a necessity for the winter camper. The best (and most expensive: $75–$150 average) bags are filled with down, which is lighter than polyester fiber insulating materials (used in sleeping bags costing about $45–$75), and can be compressed into a smaller volume—both crucial considerations for campers, who must pack their sleeping bags around with them. Sleeping bags are "rated" by their manufacturers to provide comfortable sleeping at certain minimum temperatures. When you buy a sleeping bag, however, remember that all such ratings are only generalized estimates: the temperature below which you will feel cold in your sleeping bag depends on factors which have nothing to do with how the bag is made, including ground temperature, wind-chill, altitude and your own metabolism. For the occasional camper—in winter and other seasons—who does not expect to sleep out in sub-zero cold, a relatively light, down-filled bag rated to be warm at ten to fifteen degrees above zero Fahrenheit should serve for camping trips in winter, and in milder seasons as well.

Another winter camping necessity is a small white gas **camp stove** (weighing, usually, from one to four pounds with fuel and costing $20–$40). A gas stove can be vital on trips when fire is needed in a hurry and fuel wood is scarce or hidden under snow. Make sure the stove you buy will perform well in cold weather (stoves burning butane usually will not).

For shelter on a winter camping trip you can take a tent, or, if there is a good snow cover, you can fashion a **snow shelter**. To make a snow shelter, dig in the snow a three-foot-deep trench, eight or nine feet long and as wide as it needs to be to accommodate comfortably the membership of your expedition. Don't dig down to earth: leave at least six inches of packed snow at the bottom of the trench. Line the bottom with a sheet of polyethylene, and six inches of hay, straw, dry leaves or needles on top of the polyethylene for insulation (if you have no polyethylene liner, use a foot of hay or leaves). Roof three quarters of the trench over with thin boards placed crosswise to support a canvas or polyethylene cover, weighted down with snow. The roof of a trench shelter is at ground level, so it is necessary to mark its four corners with poles or flags lest it be covered with snow and lost. Your cooking fire is built in the trench before the roofed-over section (in any winter camping situation fires must be built in a pit or trench, or they will simply sink away into the snow). A properly constructed snow trench shelter with a sturdy roof can be used over an entire winter. In constructing and using a snow trench, take care to keep snow out of the sleeping area. It will melt, wetting sleeping bags and straw, and reducing the warmth they provide.

If you prefer to shelter in a **tent**, get one with double-wall construction, having an inner liner, separated by an air space from the exterior. This arrangement prevents water condensation from forming on the inside of the tent when the outside temperature is low. By all means get a tent with a sturdy floor which is continued up the tent sides for a few inches. Nylon has replaced cotton as the preferred tent material: it is lighter and more waterproof than cotton, and won't mildew if packed damp. Most two-or-more-man, winter-worthy tents are expensive ($100 and up); rental or group purchase of a tent may be worth investigating for the occasional winter camper.

In pitching your tent, choose a location where it will not be loaded down with snow (if necessary build a snow-block windbreak to keep blowing snow away from the tent). Tents are not made to hold up under heavy snow loading. Again, learn to pitch your tent so the sides are taut and don't flap. Tent sides which flap in the wind act as bellows, pumping warm air out of the tent, where you want it to stay.

SNOW DOMES

Although few winter campers, perhaps, will take the time to construct a snow house or snow dome for their shelter, a backyard snow dome is a project which can add enjoyment to anybody's winter. Domed snow houses are fun (though, at first, not easy) to build, and even more fun when they are finished: children love them.

To build a snow dome, work with a helper. First pack down snow by treading back and forth over an area of about forty square feet with snowshoes or skis. You want snow which is packed to a uniform density through its entire depth, for you will cut building blocks for your dome out of it. Test the snow for the proper degree of packing by pushing a stick or inverted ski pole down through it: the pole should run smoothly into the snow, and not encounter uneven resistance from buried ice strata.

When the snow is ready, describe a circle of about ten-foot diameter near by, in the area you have selected for your dome. Dig into the packed snow and begin cutting blocks in vertical sections. The blocks should be about thirty inches long, by eighteen inches wide, by six inches thick. Cut blocks with an old hand saw, machete or similar broad implement. Lay the first course of blocks end-to-end around your ten-foot circle, trimming ends to achieve close fits. Lay blocks of the second course directly on top of the first. In successive courses, "break joints," or lay blocks so their ends don't line up with the ends of the blocks below them.

In building a snow dome, one man cuts blocks and the other stands inside the dome, adding blocks to the walls. The walls are supposed to rise in a spiral, and also to converge at the top, creating the dome, Therefore, you must cut blocks on a slight diagonal from end to end so the wall tends upward as successive courses are laid. You must also trim the inside bottom edges of blocks so they lean in at the top, producing the curved sides of the dome.

As the dome begins to close over the house it will be difficult

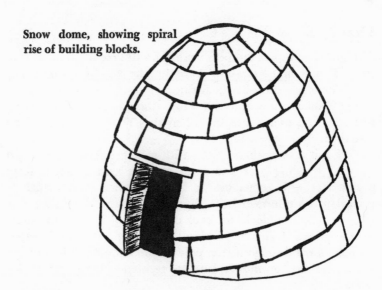

Snow dome, showing spiral rise of building blocks.

to pass blocks across the wall. Cut a temporary door in the house just large enough to allow blocks to be slid in. Cut the last block a bit larger than the hole at the top of the dome, lift it through the hole edgewise from inside, and then lower it into place, cutting it to a taper so it seats snugly.

When the dome is finished, shovel snow against the outside to seal cracks. You can cut an entrance in the wall bracing the top with a board lintel if necessary, or, if the snow is deep, you can tunnel from inside the house under the wall to the outside. Keep the entrance door or tunnel opening on the lee side of the house.

In occupying your snow dome, remember to insure ventilation. All types of winter shelter—but especially snow houses, which can be made relatively tight—present a danger of carbon monoxide poisoning. Keep the doorway open, and remove the top block to vent the roof of the dome, especially if you have a stove or candle inside.

Snow domes can be put up in an hour or two by an experienced builder, and will last for about a month.

Cross-country skiers, winter campers, snowshoers, snowmobilers and amateur snow architects are not the only ones who enjoy winter, to the benefit of their minds, spirits and bodies. Other winter activities which can help get you outdoors and have fun are skating, curling, iceboating, icefishing, following animal tracks, photography and winter bird-watching (aided by outdoor feeders). Some winter exploits really need a crowd. Get together with like-minded friends and neighbors and organize a sleigh-ride, a cross-country "citizens' race," a mini-Winter Olympics or winter carnival. In a sense, the real limits to winter enjoyment are imposed, not by weather, but only by your own imagination and invention.

GOOD READING FOR CABIN
FEVER AVOIDANCE

Brandreth, Gyles. *Games for Rains, Planes and Trains.* 1976. The Stephen Greene Press, Brattleboro, Vermont. Available in bookstores or from the publisher for $4.25 (paperback).

Brandreth, Gyles. *Home Entertainment for All the Family.* 1977. The Stephen Greene Press, Brattleboro, Vermont. Available in bookstores or from the publisher for $5.95 (paperback).

Both Brandreth titles contain hundreds of novel game ideas, with puzzles, quizzes and songs.

Appendix I *Converting Temperatures in Fahrenheit to Celsius and Vice Versa*

Fahrenheit	Celsius		Fahrenheit	Celsius
−40	−40.0	(freezing)	32	0
−35	−37.2		35	1.7
−30	−34.4		40	4.4
−25	−31.7		45	7.1
−20	−28.9		50	10.0
−15	−26.1		55	12.6
−10	−23.3		60	15.5
−5	−20.6		65	18.1
0	−17.8		70	21.0
5	−15.0		75	23.6
10	−12.2		80	26.6
15	−9.4		85	29.1
20	−6.7		90	32.2
25	−3.9		95	34.6
30	−1.1		100	37.8

To convert degrees Fahrenheit to Celsius (Centigrade) subtract 32 from Fahrenheit reading and multiply remainder by .55.

To convert degrees Celsius to Fahrenheit: multiply Celsius reading by .180 and add 32.

Appendix II *On Wind-Chill*

When we speak of wind-chill, or the wind-chill effect, we refer to the fact that cold air feels colder when a wind is blowing. In particular, wind-chill temperature equivalence tables have been developed from experiments with human subjects exposed to given temperatures at given wind speeds. Figures expressing the effect of wind-chill in these tables are based on measurements of the rate of heat loss from subjects' bodies. The tables show, then, that heat loss from bodies at a temperature of twenty degrees Fahrenheit in a ten-mile-per-hour wind is similar to heat loss at minus six degrees with a three-mile-per-hour wind (that is, calm).

It is important to appreciate that temperature equivalences in wind-

Standard Wind Chill Table Showing Equivalent Temperatures Produced by Wind at Various Speeds at Various Temperatures

Wind Speed (miles per hour)	Actual Air Temperature (Degrees Fahrenheit)					
Calm	35	30	25	20	15
5	33	27	21	16	12
10	21	16	9	2	−2
15	16	11	1	−6	−11
20	12	3	−4	−9	−17
25	7	0	−7	−15	−22
30	5	−2	−11	−18	−26
35	3	−4	−13	−20	−27
40	1	−4	−15	−22	−29
45	1	−6	−17	−24	−31
50	0	−7	−17	−24	−31

chill tables do not always give an accurate account of reality in practice. Subjects in the wind-chill experiments were shaded and unclothed; heat loss from their unprotected bodies was more rapid than it would be from the body of an individual who was normally clothed for winter. Therefore, the real effect of wind-chill on a warmly dressed person would not be as extreme as the tables suggest. Keep in mind, too, that wind-chill tables give estimates of the wind-chill effect on the human body, and cannot be directly applied to other substances (the wind-chill effect on your water pipes, for example, will be different from the effect as recorded on the standard wind-chill tables) .

10	5	0	−5	−10	−15	−20
7	1	−6	−11	−15	−20	−26
−9	−15	−22	−27	−31	−38	−45
−18	−25	−33	−40	−45	−51	−60
−24	−32	−40	−46	−52	−60	−68
−29	−37	−45	−52	−58	−67	−75
−33	−41	−49	−56	−63	−70	−78
−35	−43	−52	−60	−67	−72	−83
−36	−45	−54	−62	−69	−76	−87
−38	−46	−54	−63	−70	−78	−87
−38	−47	−56	−63	−70	−79	−88

Index